TURNING SCIENCE INTO THINGS PEOPLE NEED

Voices of Scientists Working in Industry

By
David M. Giltner, Ph.D.

**PREVIEW EDITION
(EPISODES 1-10)**

Turning Science into Things People Need: Voices of Scientists Working in Industry (Preview Edition)
Copyright© 2010 David Giltner
http://scientists.50interviews.com

ISBN # 978-1-935689-04-1

Published by
Wise Media Group
444 17th Street, Suite 507
Denver, CO 80202
www.WiseMediaGroup.com

WISE
MEDIA GROUP

Based on original *50 Interviews* concept by Brian Schwartz.

Preview edition.
Printed in the United States of America.

ACKNOWLEDGEMENTS

I am very grateful to all of the scientists who appear in this book and graciously volunteered their time to tell their stories. Their contributions have been invaluable and I have learned much from speaking with them.

Two of these scientists in particular made additional contributions to the project. Dr. Roger McGowan, my lab partner in grad school and good friend ever since, helped me distill my thoughts into a coherent series of interview questions. Dr. Jason Ensher provided valuable input that helped me craft the key messages behind the project. For their contributions, I am very grateful.

My grad school advisor, Dr. Siu Au Lee is a very important part of this project, as graduate school is where I truly learned how to be a scientist.

I want to thank my mom, Julie Giltner, for her never-ending support and encouragement.

I inherited my innate curiosity and technical aptitude from my late father, John Giltner. The love of learning he instilled in me was a strong driver for this project.

Finally, I want to thank Brian Schwartz for having the vision and the dedication to create this opportunity. The 50 Interviews concept is a powerful way to collect insight from others and synthesize an effective message.

PREFACE

Working in industry is often viewed as a nontraditional career choice for a scientist. If you are a student working towards an advanced degree in one of the sciences, or if you have recently finished graduate school and are early in your career, you may see industry as a less than desirable career path.

The purpose of this book is to change your mind.

Not by telling you that tenured academic positions are available only to a small number of Ph.D. qualified scientists. You probably know that.

Not by reminding you that nearly every professor you knew in graduate school has spent their entire career in academia, and are therefore biased in their worldview. You probably are aware of that as well.

The purpose of this book is to change your mind by introducing you to a number of scientists who have built rewarding careers in industry, and would make the same choice again, if they had it to do over.

From the detail-oriented engineer who strives to improve product reliability, to the visionary entrepreneur who wishes to change the world, I will introduce you to scientists who work in a wide range of roles in industry and have a great time doing it. I'll also give you the option of envisioning yourself in your own rewarding career in industry, and the confidence to begin moving in that direction today.

This book will show you that the 'nontraditional' path has been a great choice for many scientists, and can also be a great choice for you!

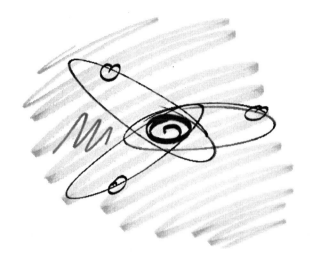

"Your work is going to fill a large part of your life, and the only way to be truly satisfied is to do what you believe is great work, and the only way to do great work is to love what you do. If you haven't found it yet, keep looking, and don't settle."

- Steve Jobs, Stanford commencement address

TABLE OF CONTENTS

INTRODUCTION

For most of my graduate school career, I intended to follow the 'traditional' path into academia. It was only during the last year, while completing my dissertation research, that I decided a career in industry would be a better fit for my skills and interests. Rather than spend years working to achieve the coveted position of a tenured professor, I decided I preferred the fast paced world of product development, where projects come and go, and results must be achieved within a matter of weeks or months.

I loved the excitement of discovery that is central to science research. I loved the challenge of solving complicated problems and the satisfaction that comes from learning something new every day. However, I didn't like the fact that it might be twenty years before my fundamental science research finally found a practical application. I did not like the fact that only a few hundred people around the world understood the work I was doing and why it was important. I wanted a much faster return on my investment. I wanted to see tangible results from my work that would help people today, or at least within a year or two.

I wanted to turn science into things people need.

My fellow graduate student and lab partner, Roger McGowan, also wanted a career in industry. He and I discussed this issue at length during our final year in grad school. We knew the challenges that lay before us. Our physics training had prepared us very well for a career in science research, but we had absolutely no idea what awaited us in the commercial world. Product development is largely about design, and we had no specific design skills. We had spent many hours working in the machine shop building custom hardware for our experiments, but our knowledge of mechanical design and geometric dimensioning and tolerancing was still very limited. We had built analog control circuits and tweaked them until they met our needs, but we were far from expert circuit designers. We had assembled countless arrangements of mirrors and lenses and optical fibers for laser beam handling and delivery, but we had neither the experience nor the tools of a trained optical engineer.

We had no idea where our broad technical skill set might fit into a product design environment, plus we had no industry connections to help answer our questions. It was clear that if we wanted to pursue the industrial career path, we were on our own. We embraced the challenge, found the information we needed, made the connections, and built rewarding careers for ourselves in the private sector.

Today, after spending a decade and a half working in industry, my perspective is very different than when I first emerged from graduate school. I now know that a broad technical skill set is an ideal background for working as a system engineer or leading a product development team. I am much better at selling myself as a skilled problem solver who is not afraid of technologies I've never worked with before. I know I can quickly learn whatever I need to be successful. I am also more confident than ever that I chose the right career path.

In April of 2008, I was invited to present a talk to the Physics Department at Colorado State University, where I attended graduate school. The title of my talk was 'Mapping a Physics Education into a Career in Product Development.' I spoke about the skills I had developed during my school years, and how they had been particularly useful in my career. I also discussed some of the new skills and habits I needed to develop in order to be successful. I spoke about how rewarding my career had been and specifically what aspects of my job appealed to me most.

My talk was well received by the attending students. This renewed my interest in career development for scientists working in industry. I began to track down a number of my 'scientist turned engineer' friends, to ask them about their own experiences. It turned out many of them had faced similar challenges transitioning from science research into industry, but they were also quite aware of how their science training gave them a unique perspective and valuable skill set. I found they were very happy with the career paths they had chosen, and excited to tell me their stories.

It occurred to me that Roger and I could have benefited greatly from hearing these stories back when we were still graduate students. They might have bolstered our confidence and given us a significant head start in our job search. I decided to collect a number of these stories and publish them in a book, aimed at helping early career scientists who might be considering working in industry, or have already taken the plunge.

In this book, I speak to scientists who work in fields typically regarded as the domain of engineers. I ask them about the skills and attributes they have found to be most useful in industry, as well as the skills they may have lacked, and how they needed to change their approach in order to be successful. I ask them what books and references have helped them in their pursuit of a rewarding career, and how their perspective has changed since they graduated. I ask them about their most outstanding accomplishments, and what they feel has contributed most to their success.

Each interview outlines a role in industry that is well matched to a scientist, providing a selection of possible jobs for the reader to consider in their own career planning. Several of the scientists I've interviewed are successful entrepreneurs, so I asked them what skills they look for in the scientists they hire for their own companies. The data that I've collected should serve as a career management reference.

I wanted to convey more than just data with this book, however. I wanted to explore the human side of these scientists' careers; what drew them to science when they began their education, what inspired them to go into industry, and what challenges they faced in making the transition. I wanted to find out what excites them most about the work they are doing now. I wanted to know what about their jobs satisfies that innate curiosity and love of learning that is fundamental to most scientists' personalities.

Data is an essential element of any career planning activity, but it is important to remember that passion and excitement are human qualities that make life worth living and a career worth pursuing. There are already plenty of books on career management that give advice, list useful references, and give tips on writing resumes, cover letters, and so on. I wanted this book to be different. This book is about scientists and their stories.

Product development is a very different activity than science research. Both are invaluable and both offer very rewarding careers for those who enjoy the process of scientific discovery and the challenge of problem solving. My hope is that in addition to finding useful data for managing your career, you also find something in the human side of these stories that inspires you to pursue a career path that is truly rewarding, where you are doing work you love and confident that you are making a difference in the world.

Enjoy.

David M. Giltner, Ph.D.
Boulder, Colorado

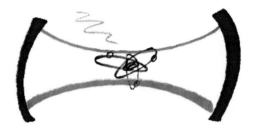

*"People are always blaming their circum-
stances for what they are. I don't believe in
circumstances. The people who get on in the
world are the people who get up and look
for the circumstances they want, and if they
can't find them, make them."*

- George Bernard Shaw

"Job growth and opportunities in the commercial science field is taking place within smaller, entrepreneurial companies, and we need to be training our graduate scientists to comfortably transition and operate within that changing environment."

- Stuart MacCormack, Ph.D.

Stuart MacCormack is a Mergers & Acquisitions Program Manager at JDS Uniphase in Milpitas, California. He has a B.Sc. and a Ph.D. in physics from The University of Southampton in the United Kingdom where he specialized in nonlinear optics. After graduate school, he continued research in this area through a postdoc and subsequent Research Faculty position at the University of Southern California (USC). After USC, he accepted a position as a Staff Scientist at SDL Inc. (later acquired by JDS Uniphase), developing high-power fiber laser technology and optical communications products. Aside from a one-year stint at the laser company Spectra Physics, Stuart has spent his entire industrial career with SDL/JDSU.

I worked with Stuart on a number of occasions during my tenure at SDL. He has always impressed me with his broad skill set. The diversity of opportunities he has had in his career is a testament to the usefulness of this breadth. Stuart's experience is a bit unique compared to the others I've interviewed for this book, because his degree is in physics, but he spent his graduate school years in an interdisciplinary group working alongside electrical engineering students. As Stuart describes it, this facilitated his transition into an industry environment.

STUART MacCORMACK

INTERVIEW

Q. You mentioned to me that you had a somewhat unique experience in graduate school at The University of Southampton. Can you describe this?

A. The Optoelectronics Research Centre at Southampton was set up in 1989. There was a strong laser team in the Physics Department and a strong optical communications team in the Electronics and Computer Science Department, so they brought those two teams together as an interdisciplinary research center. This removed the somewhat artificial barrier between what was a physics topic and what was an electrical engineering topic in optoelectronics.

Q. Can you give me your thoughts on the differences between science and engineering?

A. The goal of the scientist is creating and refining a model which describes how some aspect of the universe functions. To an engineer, the scientist's model is a tool they use to achieve an end goal, which is typically to develop some kind of device or functionality to address a real world need. For example, in optoelectronics the scientist may study the nuances of a three-level transition in erbium doped glass and try to understand the relaxation coefficient or the Einstein A and B coefficients in an attempt to refine the model. The engineer is concerned with how to utilize this model to get the best noise figure, gain, or pump efficiency in an erbium fiber amplifier.

> *"I've always worked with the philosophy that if you do the best job you can on whatever task you are asked to work on, then opportunities will open up for you."*

Q. So in a cross-functional department you have scientists and engineers working side by side, refining the model at the same time it is being utilized to develop the technology? It seems that would give you a great flexibility and ability to

2

optimize the approach.

A. Absolutely. Improvements to the tool are immediately com-municated to the folks who want to use that tool. You have meetings on a daily basis to discuss what is being learned, and how that can be applied to the real world engineering.

Q. That must help the physics students develop a stronger ap-preciation of engineering and vice versa.

A. Yes, it did. Optoelectronics is a unique field where there is a blurring of the boundary between what is a physicist and what is an electrical engineer. I remember sitting through a thesis defense for an electrical engineering student who was studying second harmonic generation in centro-symmetric materials and glasses. The amount of quantum mechanics in that presentation was phenomenal. It would have im-pressed the most hard-core quantum physicist. If you look at other engineering disciplines such as mechanical engineer-ing or true electronics design, those folks probably are much more specialized and more engineering focused.

Q. How and why did you transition into a job in industry?

A. When I graduated from Southampton, I had an offer for a postdoc position at USC. I thought it would be a great time and look great on my resume, so I took it. I was absolutely heading down the path towards an academic position. How-ever, soon after I started at USC, I visited SDL Inc. up in San Jose to give a seminar on my research to the R&D team and see if there was any interest in collaborating. I was very im-pressed with the level of research that was going on there, and I thought I could probably perform more productive and meaningful research at SDL than I could as a faculty member at USC. There was better research going on SDL in the fields of high power lasers and high power diode technology than there was at any university department in the entire world at that time. I realized I could continue down the academic path and write grant after grant, hoping they get funded, or I could go work for SDL, where they would pay me more to

STUART MACCORMACK

3

do similar research with better facilities and resources at my disposal. I would also get to work with very intelligent and focused people. Looking back now it seems like a natural progression, though it took me four years to finally make the jump.

Q. From my perspective, as a Ph.D. physicist who also worked at SDL, the research you guys in the R&D group did was fairly applied, not fundamental research. What is your perspective on this?

A. There was always a deliverable at the end of the day, usually for an SBIR contract, so it was not focused on developing a model like we discussed as the goal of science research. However, we weren't developing technology for any specific product either. At one point we developed a 100-watt fiber laser just to demonstrate an output power ten times higher than anyone had ever done before. If anyone had ever asked, 'Why do you want to do this?' the answer would have been, 'Because we can.' However, stretching the limits of the state of the art technology pushes the boundaries of the physical models, and often leads to new insight and developments on the underlying physics.

Q. It sounds like the structure of your academic training facilitated your transition into industry. Were there any aspects of the transition that were a challenge?

A. The most abrupt transition was when SDL began growing rapidly as telecom took off. Their stock was going crazy, and management decided to close down the R&D organization. We were all pushed out into the business units to "go develop products." That was the most disruptive experience that I've gone through in my career. The R&D group had been the heart of SDL and was the primary reason I joined the company. It was a big change to be moved from a research organization into an organization that was very business-focused.

The company was sensitive to the career disruption taking place, and encouraged several of the R&D staff to participate in a three day class at Stanford Business School called, "Market Strategy for Technology-based Companies." The core message of that class was, "A business in hyper- growth mode should focus all resources on near-term market growth objectives, even at the cost of sacrificing longer term R&D goals. If you can win enough market share, then you can go and buy the smaller companies that are still focused on R&D." The fact that the restructuring at SDL was supported by good economic principles greatly helped my transition to the more product- focused work environment.

Q. That's very interesting. So, the solution to the most abrupt change in your career was simply gaining a different perspective, and that change happened within the space of a three-day class.

A. Physicists tend to be very rational people and want to understand the root cause behind why decisions or changes are made. It was a decision that was outside of our control, so if we could get a good rational explanation, we could suddenly come away from it thinking, 'Okay, it is something I need to get used to, because obviously it makes sense.'

Q. How has your perspective changed since you first got into industry?

A. I started off with a bit of an elitist view that the problems solved by physicists were much more fundamental and important than problems solved by engineers. I absolutely do not have that view now. You can find problems that are worthwhile and extremely challenging in both academia and industry, although they may be challenging and worthwhile for different reasons.

Through my exposure to the non-technical side of the company, I've also gained a lot more appreciation and respect for other functions within the business such as marketing, sales, and finance. Running a business is a very challenging problem that requires complex strategies encompassing much

STUART MACCORMACK

more than just engineering.

Q. I also used to have the view that science is somehow more elegant and sophisticated than engineering or business. The fact is, this perspective is perpetuated in graduate school. There is an underlying attitude that everyone should want to be a professor, and that going into industry represents some sort of failure.

A. Yes, as though you've given up. Science is too hard for you, so just go and build stuff.

Q. Yes. At least we now know better. So, how do you sell yourself to an employer?

A. I've always worked with the philosophy that if you do the best job you can on whatever task you are asked to work on, then opportunities will open up for you. If you can establish a reputation as someone who successfully completes their current goals and objectives, you quickly rise through the list of candidates as new opportunities open up within your company or wider network of colleagues. I don't over-think how I should be positioning myself and how to optimize my career path.

Q. That's very interesting.

A. Now I'm thinking, "Is this why I've ended up where I am today?" [Laughs] Just think about how good my life could be if I were more forthright in thinking it through. Seriously, though, I've been very fortunate in my career in that new opportunities have come up almost yearly, and each has been a chance to learn about a new technology or a new function within the business. That is something that is very hard to achieve through a conscious career planning exercise.

Q. You have had some great opportunities to do exciting new things within JDSU. Can you describe some of these?

A. One good example was getting to work on the Critical Accounts Team. This was a small team of project management 'fire-fighters' created by the CEO with the directive of engaging with a project team or business process that was failing to achieve its goals. Our objective was to work with the existing team to develop a recovery strategy that would get the project back onto the right trajectory. This group would tackle problems in nearly every area of the company, covering several business units and many different product lines. Our high visibility to

> "The most rewarding part is the continued variety of things I get to work on, never getting to the point of being pigeonholed in a particular area."

the CEO and executive team allowed us a great deal of flexibility, and gave us access to resources from other projects and different parts of the company. This was a great learning opportunity and very fun, but it is the type of job that you cannot train or plan for. You just have to be in the right place at the right time and stumble upon it.

Another example was being asked to be the program lead for the divestiture of our manufacturing facility in Shenzhen, China. At the time, this facility employed approximately 2,000 staff. This was a major initiative for the company. The repercussions of failure would have been significant to the company, customers and all management teams involved. Had they advertised for a Divestiture Program Manager, I would never have applied, since I had absolutely no direct experience with a project of this type. However, the management team had developed sufficient confidence to offer me the role of managing this project. In this case, they were prepared to overlook the lack of direct experience, and based their decision on a prior demonstration of the more generic program management and organizational skills. They were looking for someone with good program management skills who understood the technology and the people involved, and they knew that I had been successful on the

STUART MACCORMACK

7

Critical Accounts Team. It makes a huge difference just having established yourself with your peers. People look at you as more than just a resume and a set of skills on a piece of paper.

During the last year, I've been working with the Mergers and Acquisitions Team, exploring companies we may want to purchase. It's a very fun environment with lots of new things to learn. The most rewarding part is the continued variety of things I get to work on, never getting to the point of being pigeonholed in a particular area. The price you pay, is that you end up becoming a generalist and know a little bit about a lot of things, but give up being a functional expert in any one area.

Q. You've spent the last 14 years of your career here in Silicon Valley, and most of that in one company where you built a solid network of people who know your work. Do you think that has helped you find good opportunities without having to actively plan and sell yourself?

A. Yes, I do. From a career perspective being a generalist is very rewarding, but it's very tough to sell yourself. If I ever go look for a job at another company I will have to sell myself as a jack-of-all-trades, but a master of none. Silicon Valley is unique, however, because there are just so many companies in a small region and a cross-fertilization from people who leave companies but stay in the area. I guess it has got to be more challenging when you pick up and move to a different area.

Q. Where do you see yourself in ten years?

A. I find it nearly impossible to extrapolate from my current trajectory and guess where my career is going to be ten years from now, but I would prefer not to be working full-time and achieve a better work-life balance than the typical Silicon Valley technologist. I'd like to have enough of a demand on my skill set that I can engage in consulting work as required.

M & A PROGRAM MANAGER

Staying in my current role in mergers and acquisitions, or similar project management roles, should allow me to develop a very marketable skill set and good network that I can take advantage of in the future. These will be very marketable skills here in Silicon Valley.

Q. Anything else you would like to add?

A. The topic of career paths for graduate scientists is very timely, since the era of the large corporate research labs is nearing its end. The option to follow a science career within corporate America is becoming an unrealistic option for new science graduates. Job growth and opportunities in the commercial science field is taking place within smaller, entrepreneurial companies, and we need to be training our graduate scientists to comfortably transition and operate within that changing environment.

STUART MACCORMACK

"Control your own destiny or someone else will."

\- Jack Welch

2

"It is important to communicate your successes without appearing arrogant, to be confident in your abilities, but also be willing to admit when you don't know the answer."

- Roger McGowan, Ph.D.

Roger McGowan is a Research Fellow at Boston Scientific in Maple Grove, Minnesota, where he develops fabrication processes for vascular stent delivery systems. He received a Ph.D. in physics from Colorado State University, performing experiments in laser spectroscopy and laser light manipulation of atoms with Dr. Siu Au Lee. After receiving his Ph.D., Roger took a post-doc position with Dr. Dan Grishkowski at Oklahoma State University. He then returned to the Minneapolis/St Paul area where he grew up, working at Imation and ADC Telecommunications before arriving at his current position with Boston Scientific.

In high school Roger was a self-proclaimed "gear head" and says the interest in mechanical design and operation he gained while working on his 1970 Dodge Challenger influenced his professional life. He also credits his extensive work with youth activities in college with helping him understand how to work with people, a skill that he has found very useful in his career.

I first met Roger when we began graduate school together in 1989. Through many long hours working in the same research lab we became close friends and I am very pleased to include his contribution to this project. I've always been impressed by Roger's strong focus on interpersonal communication. It is no secret that scientists and engineers are not typically known for their strong social skills, but Roger has demonstrated that these are a valuable component of career success.

ROGER MCGOWAN, PH.D.

INTERVIEW

Q. Tell me about the science phase of your career. What originally attracted you to a science education?

A. My interest in science was primarily driven by a desire to understand how things work. I've enjoyed science since the fourth grade, but I didn't really become a scientist until graduate school. Once I got into the lab, I was faced with having to rely on my own understanding because the answers were no longer available in a textbook. That's how a real scientist works.

Q. Has there been anyone who served as a role model or mentor for you?

A. My advisor in graduate school was certainly a mentor. She excelled in giving me an interesting opportunity and then challenging me to persevere and achieve great results. She held the bar pretty high.

Q. What motivated you to pursue a career in industry after your post-doc?

A. By the end of my post-doc I knew I wasn't interested in the lifestyle of a university professor. It's hard to achieve a good work-life balance when you have to be an instructor, a researcher, and a proposal writer all at the same time.

I knew that I would like to teach someday, but I wanted to have a solid background in industry first. Many physics students will end up in industry, yet most professors don't have an industrial background. I feel that if I want to be a really good teacher, I should first have that industry experience so I can help my students make the transition.

> *"Physicists make very good technical leads and project managers because of their critical analysis and problem solving skills."*

Another reason I went

RESEARCH FELLOW - PROCESS DEVELOPMENT

into industry is money. I knew I was going to raise a family and wanted the higher salary typically found in industry. Finally, I liked the idea of creating something useful.

Q. Do you have any thoughts about the typical differences between scientists and engineers?

A. I try to be careful not to stereotype. However, after many years of working with both scientists and engineers, one thing I notice is that scientists tend to be much more comfortable asking 'why' and 'what' questions in pursuit of a root cause. As an example, suppose a fabrication process stops working when we start using a new batch of a certain material. I've found that engineers are more likely to grab another batch of material and try to get the process running properly again. A scientist is more likely to try to figure out what property of the material changed and why, knowing that this will allow them to better control the process in the future. It comes down to digging deeper to achieve a scientific understanding of the problem versus simply adjusting the parameters until a robust process is achieved.

Another key difference is innate curiosity. A desire to understand what's going on at a fundamental level led me to become a scientist rather than an engineer. In engineering you may learn about the tools and how to use them, but you don't necessarily learn the fundamental science behind them.

Q. What is one of your biggest achievements in your career to date?

A. When I first joined Boston Scientific I worked in a process development team that was designing a new laser bonding process for their stent delivery catheters. The existing process had been optimized by varying parameters until the desired result was achieved, but the team didn't understand why it made a good catheter. The new process I helped develop had many controls and monitors so we could observe

ROGER MCGOWAN, PH.D.

the effect of varying each parameter as we were making a catheter. It was essentially an engineering project, but what made it truly successful was my emphasis on a fundamental scientific understanding of the laser weld.

This project was successful enough that the new process was extended to all of our next-generation laser systems. Championing this effort led to a lot of career development for me within the company. Showing them the value of a scientific approach led to the creation of my current position, Research Fellow for Process Development. Being successful in this effort is what I'm most proud of in my career.

Q. That is a great demonstration of the benefit of using a scientific approach in a product design setting. What would you say are the top two or three skills that enabled this achievement?

A. The most important thing was a solid technical understanding and the ability to teach myself. These are skills that I learned in graduate school. They gave me the confidence to jump in and quickly learn about CO_2 lasers and other things that were new to me. Without these skills, I wouldn't have had the ability to demonstrate how my ideas would improve the process.

Another important skill was my experience in teaching and mentoring. Much of my success resulted from presentations that impressed senior management. It was critical to be able to teach them and help them understand why my ideas were of value.

My ability to dissect a problem was also very important. Any technology brought to market will have numerous challenges along the way. You have to be efficient and effective at leading a team to dig in and work through these problems.

Q. Do you have any other thoughts about ways that a science education might prepare someone for a career in industry?

RESEARCH FELLOW - PROCESS DEVELOPMENT

A. It is important to understand data and the factors that might influence it, including the measurement system itself. The pace in industry is very fast and it is tempting to just accept your data and move on. I find that if I dig into how the data was collected and what equipment was used it is often not so simple or so convincing. I've found that when you show people a graph they tend to accept it as fact, but just because you have a pretty graph doesn't mean it's correct.

Q. Are there any specific roles in an industrial environment that you feel are a great match for a physics background?

A. A scientist who is right out of school can fit into an individual contributor role in almost any industry, particularly in a company with a strong R&D focus. The options increase as one gains experience. Physicists make very good technical leads and project managers because of their critical analysis and problem solving skills. They're able to ask the right questions and understand many aspects of a project simply because they have a basic understanding of a lot of different areas.

Process Development is a good fit for a scientist because there's a lot of problem solving involved. Design Assurance and Quality Assurance involve a lot of experimentation and statistical analysis and are also a good fit. There's also Sustaining Engineering, which involves making changes or improvements to an existing product. A science background fits all of these roles.

It's very important to make students aware of these opportunities in industry. I had no idea that these types of positions even existed when I was coming out of school.

Q. Neither did I. How do you sell yourself and your skills to a potential employer?

A. First, I've got technical vision and leadership skills. I can see how to make products better and then lead a team to achieve

ROGER McGOWAN, PH.D.

15

this vision. Second, I have a broad technical background. That allows me to dig in to any aspect of any product and work through any problem. Third, I am skilled in communication, teaching, mentoring, and public speaking – all of which are critical for selling a project inside and outside of the company. Fourth, I am skilled at problem solving, root cause analysis, and statistical analysis, all of which are essential for creating a robust product.

> *"A huge part of my success in industry has come from my communication skills and my focus on developing good working relationships with my colleagues."*

Q. What have been some of the aspects of your transition to industry that were a challenge? Are there areas where you didn't feel you were well prepared?

A. One challenge was investing several years in a Ph.D. and finding that it didn't count as 'experience' from an industry viewpoint. I also had to shift my mindset from pursuing projects that were interesting to identifying what would generate revenue. In industry all eyes are focused on revenue generation, but this is not the purpose of science research.

There are companies where you can walk in with a science background and feel right at home. Some larger companies still have research labs, and there are smaller companies focused on developing a core technology that may offer a lot of R&D play-time. However, many companies are very focused on production, yields, and quality, and that creates a very different environment. I suggest that a job seeker ask questions when they interview to help identify what kind of company they're looking at.

A bit of technical knowledge I never learned about before working in industry is Design of Experiments (DOE). I was also lacking a solid understanding of the statistical techniques used to study reliability of a product or repeatability

RESEARCH FELLOW - PROCESS DEVELOPMENT

of a manufacturing process. In physics research you are not trying to create a product that will last a long time or can be replicated in large volume, so these topics are not typically covered in school. It also would have helped to have an understanding of the basic stages of product development and design verification processes.

> "It's OK to make mistakes. We just have to make them faster than our competition."

Q. **How has your perspective changed since you first went into industry? Is there anything you would do differently with the knowledge you have today?**

A. One thing I would have done differently is to plan for going into industry while I was in grad school. I value what I learned from my post-doc, but I would skip it if I had it to do over again. Now I realize that three years gaining experience at a company would have been far more meaningful than the three years I spent as a post-doc.

When I first went into industry I envisioned myself ultimately reaching a director-level position and guiding a team to achieve a common vision for our technology. My experience since then has shown me that a Director of Technology doesn't necessarily deal much with technology. I wouldn't want a director level job at the large company I'm with now because they spend their time pushing paper and dealing with budgets.

Q. **What have been the most rewarding aspects of your career to date? What excites you or keeps you motivated in your career?**

A. I enjoy working in the medical industry. Making devices to help people and save lives can be very rewarding. I also enjoy the mentoring and the teaching opportunities that come with leading teams and projects.

Roger McGowan, Ph.D.

As for what keeps me motivated, that would be a desire to create something and to have an impact on the company. I have a strong internal drive to develop myself, exceed the expectations of my manager and team members, and consistently achieve larger and larger opportunities.

Q. Where do you see yourself in 10 years?

A. I see myself in a smaller company or perhaps a startup. I still have that burning desire to find that great idea and go turn it into something real. I want to take all these experiences I've gained to a smaller company where I can have a larger ownership and responsibility and see things move faster than they do in a big company.

Q. Do you have a favorite book or resource that you recommend?

A. I don't have any specific favorite, but one that I enjoyed because of its perspective on scientific thinking is <u>Surely You're Joking, Mr. Feynman!</u> I thoroughly enjoyed reading that book because many of his thoughts and his interests resonated with my own. It was enjoyable to find out that someone who made such an impact on the world wasn't that much different than the rest of us.

Q. Are there any particular sayings, quotes, mottos, anything that you live by?

A. A very important perspective that I learned from my advisor and use when talking to my team is "It's OK to make mistakes. We just have to make them faster than our competition."

Q. Is there anything else you'd like to share with a fellow scientist who's considering a career in industry?

A. A huge part of my success in industry has come from my communication skills and my focus on developing good working relationships with my colleagues. It is important to communicate your successes without appearing arrogant, to be

RESEARCH FELLOW - PROCESS DEVELOPMENT

confident in your abilities, but also be willing to admit when you don't know the answer. I call this approach 'humble self-promotion' and it summarizes much of my attitude on how to live your life and build your career. I believe the reason my work associates often become good friends is because I'm genuine with them.

Roger McGowan, Ph.D.

"Scientists, like hangmen, are socially disad-
vantaged by their profession."

- Len Fischer in <u>How to Dunk a Doughnut:</u>
<u>The Science of Everyday Life</u>

"The rush that helps drive me is making a sale and shipping something out that was unbelievably hard. It's about being able to shape the world, and that's very cool."

- Chris Myatt, Ph.D.

Chris Myatt is the CEO of Precision Photonics Corporation (PPC) in Boulder, Colorado, a company he founded with his wife Sally Hatcher in 2000. PPC develops precision optical components, coatings and assemblies for the telecommunications, aerospace and defense, biomedical, and semiconductor industries. He also guided the incubation of mBio Diagnostics, a subsidiary of PPC that develops low-cost medical devices.

Chris earned a B.S. in physics and a B.A. in mathematics from Southern Methodist University, and a Ph.D. in atomic physics from the University of Colorado. As a graduate student, he worked with Carl Weiman on the Bose-Einstein Condensation (BEC) project, for which Carl received the Nobel Prize in physics in 2001. After graduate school, he went on to a postdoctoral fellowship at NIST in Boulder, working on ion trapping. In his spare time Chris enjoys cycling and hockey.

INTERVIEW

Q. Please tell me about the science phase of your career.

A. I have always liked math and science. They were the easiest for me. I received an undergraduate scholarship to Southern Methodist University, where I got hooked on lab work, and decided that's what I wanted to do. I was an undergraduate and wasn't given a lab key, so when the building was locked up in the evening, I would leave through the door, go get a Big Gulp from 7-11, and come back through the window and keep going until three or four in the morning. We were doing a laser spectroscopy experiment, and my primary contribution was debugging a problem with a noisy power supply. I ended up publishing a paper on that project.

The scholarship to SMU included a trip to Europe for a summer. I went to Austria and decided I liked mountains, so I applied to graduate schools near mountains. CU-Boulder had mountains, and it also had professors who were prominent in laser spectroscopy, so that's where I went. While at CU, I worked on Carl Weiman's BEC experiment and demonstrated the first two-state BEC system. That's what my Ph.D. thesis was on.

Originally I wanted to be a professor, because I like teaching and I love explaining things. It's always fun figuring out how to help someone else understand what you are talking about. What I hated was writing papers. The academic world is structured so that you need to write a paper at the end of each project. That was a huge turn-off to me. I realized that this wasn't going to work during my postdoc, so I went and found an industrial job.

> *"I try to maintain a perspective on why my work would matter to my grandma. I figure if my grandma can understand my business plan, then it is probably a pretty good one."*

Q. Tell me about your transition from science research into industry.

A. In 1999, Research Electro-Optics here in Boulder was hiring like crazy to meet the telecom demand, so I went to work for them. After a year I decided that I wanted to do more than just optical components. We founded PPC to go after more complex laser frequency control projects, utilizing my background in laser control systems and diode lasers. In the end it turned out that we needed to make optical components as well, because everyone we tried to buy them from was swamped by the telecom demand. The businesses we could win was making component, and in the end that has turned out to be a very robust business.

Q. How did you make the switch to the medical industry?

A. It started in 2003 when telecom had just gone to hell. Milton Chang, one of our investors, said "You've got to go find something else to do." We deliberately sat down and thought about what would be the coolest thing that I could do as a measurement guy, combined with the skills I had developed within the company. Low-cost medical diagnostics seemed like the right thing. It fully ignites all my technical skills, it's a challenging business problem, and it is a part of the solution to a major problem in the world. It was a deliberate decision that our new project would be something that was used all over the world, because now I get to see the world while I am running my business. Those are all deliberate decisions, and it's so cool that it's working.

Q. What do you consider your biggest achievement at this point?

A. Getting PPC to where it is stable, makes money and gives us the freedom to start a medical diagnostics business has been a huge achievement. We started PPC in 2000, and were profitable in 2006. We made a number of missteps and had some major challenges in 2007 and 2008, but recovered from them. We were able to turn a profit in 2009, and should be doing very well in 2010.

Q. What are the skills that allowed you to achieve this?

A. It was not a brilliant business plan and it was not luck, because we had a lot of bad luck. The biggest thing has been perseverance. My wife and I have a lot of our money tied up in this company and so do our families. When you have that level of personal investment it's "Never Say Die." We have

CHRIS MYATT

23

had two very close encounters with death of the company and several near misses. Many others might have given up.

Another important thing is an attitude about the time you invest. When you are running your own business, there is a blurring of the boundary between work and personal time. My wife works at PPC as well and it consumes 80%-90% of our time. I learned about this at an early age from watching my dad. He was a self-employed dentist and worked a tremendous number of hours. If something went wrong with the business, it didn't matter what our plans were, it was, "Come on, we are going to go fix the plumbing in the office."

It's also very important to focus on the problem at hand and not waste time on non-critical details. My dad was a very practical thinker and I picked that up from him. This attitude worked well in graduate school with an advisor like Carl Weiman. His opinion was that if you are getting more than a B in your classes, you are not spending enough time in the lab. Every physics student wants to talk about electrodynamics and field theory, and other cool things, but his attitude was "Get back in the lab and get the experiment working." Graduate school also prepares you for the blurring of personal and work time that I mentioned before.

Once I jumped into business and realized the problem at hand was that a paying customer expected me to deliver something to them, it wasn't so hard to make that transition.

Q. How do you feel that your science training has contributed to this success?
A. The science training means that I can do this all in a technical field. It lets me play in an area where I have an advantage.

However, sometimes I consider it a disadvantage, because it makes it harder to be a good salesman. I try to maintain a perspective on why my work would matter to my grandma.

FOUNDER - PRECISION PHOTONICS CORP.

I figure if my grandma can understand my business plan, then it is probably a pretty good one. I know so many details about our technology, that it is harder to give a clear, cogent description that my grandma would understand. I can also see so many ways we could improve on what we do, but we can't make all the improvements due to other non-technical constraints.

> "Solving a problem in the real world has constraints far beyond just technology. That's only one of about 30 axes that matter."

Solving a problem in the real world has constraints far beyond just technology. That's only one of about 30 axes that matter. For mBio what really solves the problem is an adequate level of technical performance, delivered in Africa, this year, with approval from the Ministry of Health, and at a specified price point.

Q. Do you have any thoughts about the differences between scientists and engineers?

A. Scientists tend to have little connection with the applications for their technology. They may know why it matters from a scientific perspective, but not necessarily why it matters to society. An engineer is more interested in applying technical know-how to solve real world problems. They are steeped in why what they are doing matters from an industry or society point of view.

Early on I was lacking this connection to the real world as well. When we achieved BEC, the world didn't change. Most of the results were things people could sit down with a pad of paper and predict. When that finally dawned on me, I was a little bit disappointed. Around 200 academic groups have jumped on the BEC bandwagon, but so what? What will be much more satisfying to me is if I see our medical products

CHRIS MYATT

in clinics around the world in the next 10 years.

Q. Would you say that is the most rewarding aspect of your career?

A. Absolutely. The concept of being able to make a difference in the world is extremely rewarding.

Q. You employ a number of scientists here at PPC and mBio. What do you look for in the scientists you hire?

A. We hire lab rats, people who love to spend time working in the lab. Experimental physicists are lab rats and are very good at solving technically challenging problems.

We are also hiring biochemists and immunologists to develop new assays for our medical business. We need someone who can develop a new process unlike anything they have dealt with in their past career. They need to be able to translate the assays they develop with a pipette and a bunch of test tubes into a little fluidic cartridge that doesn't look like anything they are used to. When new information starts coming at them quickly, they need the ability to sort it out and say what is important and what is not.

Their science education provides an important base, but it is far from sufficient. We look for someone with a spark, flexibility, and ability to learn and process new information. When we interview candidates, we will throw problems at them and try to create a situation where they will be uncomfortable in order to see how they handle it.

Q. Is it ever a challenge to keep the scientists focused on the non-technical aspects of the problem?

A. Yes, it is. Any time you hire someone, you are going to have to teach them something, so you hire based on what you prefer to teach. I have to teach them about what matters in business and how you please your customers.

Q. Where do you see yourself in 10 years?

A. The projects that mBio is now chasing present a career's worth of challenges. I don't know exactly where that will lead us, but I would love to see that we have driven our diagnostic technology into every aspect of the medical field.

Q. Do you have any favorite books or resources that you have found useful along the way?

A. Jim Collins' books are great. My favorite is <u>Built to Last</u>. I have a pile of books at home on how to hire people, get organized, write a business plan, and finance a company. I will flip through them periodically just to pick up as many nuggets as I can.

Q. Do you have any quote or saying that you like?

A. I think everyone here gets tired of my quotes and sayings. There is a quote that I like from Steven Chu, who won a Nobel Prize in physics, developed laser tweezers for DNA, ran Berkley National Labs, and is now the U.S. Secretary of Energy. Someone once asked him what the secret was to making all these changes and doing them very successfully. He is reported to have said, "When I go into something new I make as many mistakes as fast as possible." Here is a damn smart guy who is driven like there is no tomorrow, and he has the talent to go with it. He's saying that his secret is not that he avoided more mistakes than you, but he made more than you did and learned quickly from them. That's something I don't think very many people appreciate.

I think a generalization of this concept explains why the U.S. is one of the biggest economic powers in the world. In the U.S. the mechanism for failure is pretty well-defined. Bankruptcy may not be pleasant, but it doesn't ruin your life. It allows you to come out the

> *"The concept of being able to make a difference in the world is extremely rewarding."*

CHRIS MYATT

other side and still be very successful. That makes all the difference in the amount of risk people are willing to take.

There are a lot of entrepreneurs who failed a number of times before they were successful. The Ford Motor Company that we know of was actually Henry Ford's third attempt. Too many people are afraid of screwing up, but ultimately the folks who achieved the greatest were willing to hang it out there very far.

Q. I've read several accounts from entrepreneurs who said that their best innovations were motivated by being very close to losing it all.

A. I agree. PPC probably wouldn't be where we are today if telecom hadn't gone to hell, forcing us to think quickly, move quickly, or fail.

Q. Do you have any other thoughts?

A. If you are going to go down an entrepreneurial path and be successful, you are competing with people who have blurred the lines between what is work and what is play. They are going to eat your lunch if you are not on the same level, and you have to accept that.

I remember sitting up at two in the morning after we had caught an employee stealing from the company. I was slogging through all this shit and thinking "God, I just want to go to sleep." But I realized that this is what I had chosen, and I stuck with it. At some point it's no longer a conscious conversation, it is just there. I recognize it is changing me in many ways, but I know I won't be successful if I don't accept it. What I'm trying to do is make sure I maintain a good relationship with my daughters and my wife. That's not simple - it is a challenge.

Q. Does it help that your wife is involved in the company as well?

A. She understands very clearly why I have to work crazy hours. I've noticed that many small companies are led by husband and wife teams, and it's because it is almost impossible to run a business that doesn't suck the family in, in some way.

The rush that helps drive me is making a sale and shipping something out that was unbelievably hard. It's about being able to shape the world, and that's very cool. I see that the medical industry can be very different than it is today, and we are changing that with mBio. In the future, you will walk out of your doctor's office with your test results in 15 minutes. It's going to happen, and I am going to be a part of it.

CHRIS MYATT

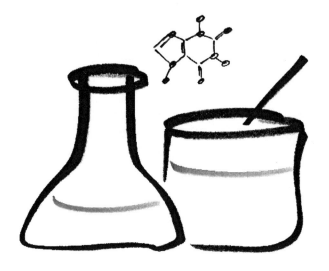

"I have always found that plans are useless, but planning is indispensable."

- Dwight D. Eisenhower

4

"If you have a good science education, then the world is your oyster, and you can do almost anything."

- Tanja Beshear, M.S.

Tanja Beshear is an ASQ certified Reliability Engineer and a Product Stability Group Leader at Covidien in Boulder, Colorado. In this role, she ensures the long-term stability and reliability of expiration dated, disposable electro-surgical devices. She received an M.S. in physics from the University of Kaiserslautern in Germany, where she focused on experimental and applied physics. She began her career developing fiber-optic transceivers at Siemens in Berlin, where she quickly found a niche in Reliability Engineering and Failure Analysis. She continued her career at Siemens, Infineon, and Osram Opto-Semiconductors in San Jose, California, before moving to Colorado to work at Picolight. In these roles, she worked as a Quality and Reliability Manager on a wide array of opto-electronic devices. She then moved into the medical device industry at Covidien.

Tanja lives with her husband and son in Lafayette, Colorado. Playing chess is a family hobby, and she enjoys volunteering as the assistant coach for the local elementary school chess team. She also finds yoga helps her stay healthy and centered.

INTERVIEW

Q. Can you tell me about the science phase of your career? What did you study and why did you originally decide to pursue a career in science?

A. I grew up in a household that was very science-oriented. My mom studied physics and math and was a schoolteacher, and my dad was a mechanical engineer in the paper industry. Growing up, we always had a copy of Scientific American and Popular Science lying around, and watched science or nature programs on TV.

I went to a technical high school in Germany with a strong emphasis on physics and math, but at first I wasn't sure what I wanted to study. Physics was very interesting, but it seemed very challenging, and there were very few women in the field. My Mom had to fight tooth and nail to get permission to study physics and math, and then she went into teaching, which was more accepted for a woman back then.

However, I was very interested in physics. I wanted to know how the universe works and what gravity is. In school you can see electricity spark and hear it sizzle, but gravity always had some mystery to it because you can't see it. I wanted to know what holds the universe together and dig a layer deeper than what we had studied in school. With the support of my parents, I went on to enroll in the physics department at the University of Kaiserslautern in Germany.

Q. Is there any particular aspect of physics that you liked?

A. I was very drawn to the applied sciences and experimental physics. My question was always: "How can this be applied? What does it mean? What can I do with this?" I was happiest in the lab turning the knobs and playing with things.

Q. Why did you decide to go into industry after you graduated?

A. The primary reason I did not continue in academia is because there were so few women in physics at the university. It wasn't until 1995 that the University of Kaiserslautern had its first female professor, and I graduated in 1989. In fact, I came across an interesting job posting this month (January 2010) that read: "The University of Kaiserslautern is endeavoring to increase the proportion of women in research and education. Qualified female scientists are especially encouraged to apply. Individuals with handicaps are preferentially

RELIABILITY ENGINEER

hired. Applicants with children are welcome." I found that interesting, that being a woman and having children is still in the same category as having a handicap. I have been out of the university system for 20 years and things have changed, but women physicists still have a ways to go in academia.

Q. Have you found industry to be a more welcoming environment as a woman?

A. Yes, but by the time I got out of school, I didn't concern myself much with the question of how I was being perceived. I had just made it as one of only two women in my graduating class in physics, so I had developed a pretty thick skin. I still remember sitting in a biology class early in my undergraduate studies when the professor walked in the first day and said there was no reason why any women should be in the classroom. He said we were not going to make it through the class anyway and it would be smarter and a better use of our time to leave right then. There were probably three women in the class and none of us stood up. We sat there and we made it through the class and got our credit. By the time I graduated I didn't worry too much about how I was perceived. I knew I was very good at what I was doing and I could do pretty much whatever I wanted. However, when I moved from Berlin to work in California, there was definitely a different climate. In the U.S. being a woman played much less of a role than it had in Germany.

> *"I wanted to know what holds the universe together and dig a layer deeper than what we had studied in school."*

Q. That's good to hear. Did you have any role models to help with the transition to industry?

A. My dad always fostered an enthusiasm for science and engineering, and that has carried over and helped me transition into engineering. One of the first things I can remember

TANJA BESHEAR

when I was really small, was my dad taking me to the paper company where he worked on Saturday mornings. I would wear a hard hat and he would show me the production line. He had a lot of enthusiasm for his work, and

> *"...women physicists still have a ways to go in academia."*

it made a deep impression on me to see the huge machines and vats stirring the solution that gets made into paper. My dad passed away before I graduated, but I already had very vivid pictures in my mind of what it meant to be an engineer or a manager, and the issues you deal with in industry.

Q. Given your experience in both science and engineering, how would you compare the two disciplines?

A. Engineers are typically more interested in building things and figuring out how they work. Scientists like taking things apart and finding out why they work. They like digging down into the fundamentals, like my question about what is gravity. It's all about not being satisfied with the formula for gravity and calculating how fast a rock falls, but wanting to dig one layer deeper and ask: "How does it actually work? Why does it work? What is the mechanism by which the earth pulls on the moon and vice-versa?" That is the scientific way of looking at things.

Early in my career I worked on failure analysis and reliability engineering of fiber optic transceivers at Siemens in Berlin. I found that if we were testing several prototypes of a new device, the engineers would be more interested in the units that worked best, wanting to fine tune them, teasing out a little bit more bandwidth and a little bit more signal to noise ratio. I was more interested in the ones that didn't work, because I liked to take them apart and find out what was wrong. That's one view of the difference. Engineers are happy when it works and they can ship it to a customer. Scientists are typically more intrigued by using a scientific approach to figure out why it works or why it doesn't.

I want to be careful not to draw a black and white picture because most of us have scientific and engineering aspects to our professional self, and if you want to develop a solid product you want to have both skill sets on the team. If you put a couple of science Ph.D.s together in a room they probably won't ever come out with a viable product, but if engineers and scientists work together on a team, you can achieve a great result.

Q. Can you tell me about a significant achievement in your career?

A. I have led a couple very challenging efforts to solve field return issues where the return rates were very low and the problems were intermittent. I get very excited about complex problems like that, and solving them gives me a lot of satisfaction.

Q. Could you summarize the key skills that enabled these achievements?

A. One important skill is something you learn very early in experimental physics, to challenge your assumptions. You tend to make certain assumptions about your products and about the customer's application for your product. To solve field return problems I had to challenge those assumptions. I traveled to the customer's site to determine for myself how their system actually worked.

An understanding of data analysis is also important. As scientists, we are trained in the skills of careful data analysis and the importance of data driven conclusions that are independent of our own pre-formulated ideas. When speed is vital it can be tempting to make a quick decision believing that we intuitively know what is happening. Our experience as scientists tells us to challenge that tendency and let the data speak for itself.

TANJA BESHEAR

My physics education has helped me a lot with failure analysis. Even an LED with one chip in a little plastic package can present a range of issues spanning electrical engineering, mechanical engineering, process engineering, and materials science. As physicists we have a solid background in all these disciplines. We also learn good study skills so we can quickly dive into a vast array of information and filter out the specific information we need.

Q. **Are there specific roles in an industrial environment that are particularly well suited to a science background?**

A. Leadership roles that require one to evaluate, synthesize, and make decisions based on a lot of different information are good roles for scientists. I have met many engineering managers, quality and reliability managers, and CEOs of startups who were scientists.

I have also seen very good software engineers who came out of a traditional physics curriculum. You can go even into finance or business management, because you develop solid math skills. It's exciting because there are not just specific niches for scientists. If you have a good science education, then the world is your oyster, and you can do almost anything.

Q. **That's a great perspective. Knowing all that you've just relayed, how do you sell yourself to a potential employer?**

A. It's important to evaluate this question regularly throughout your career. I work in reliability engineering and failure analysis, so I sell my broad technical skills. I also show that I supplement these with managerial, teamwork, and communication skills. I emphasize that I am flexible and adept at working with people from different cultures and backgrounds around the world.

I'm reminded of an old slogan for a German chemical com-

RELIABILITY ENGINEER

pany: "At BASF, we don't make a lot of the products you buy. We make a lot of the products you buy better." That's where I see myself. I haven't designed the products I work with, but I make them better. That's where I get my satisfaction from.

Q. Were there any aspects of the transition into industry that you found challenging or things that didn't come so naturally?

A. I was shocked to find how inadequate my knowledge of engineering statistics was after coming out of a physics curriculum. Surprisingly, most of my statistics training in school came from classes that were part of my biology minor, not from physics classes. My managers at Siemens were very helpful and supported me in taking classes to fill this hole.

I was no as surprised to find that I lacked management skills and presentation skills. I have friends who graduated from electrical and mechanical engineering colleges where they had presentation skill classes and technical writing classes. The curriculum I came out of had no such classes. As scientists we were taught that if we do a good job, the data would speak for itself. We thought we could present our technical paper and that would make us famous. I'm exaggerating a bit, but in industry it doesn't work that way at all.

> *"Graduation day is not the end of learning, but your first day on the job."*

Q. Is there anything else that you know now that you wish you had known or done sooner?

A. I wish I'd had a better appreciation for career management skills, how to sell yourself, how to find out where you fit best, and how to actively manage your own career. I also wish I had understood the importance of developing a strong personal brand earlier in my career.

Tanja Beshear

Q. What have been the most rewarding aspects of your career? What excites you most about your work?

A. It's important to me is that my work is never boring. Industry moves at a breathtaking pace, and there are surprises you didn't even know about on Monday that change what you had planned to do on Wednesday. Of course, Monday mornings are sometimes difficult, but in general I really enjoy going to work. The wide variety of experiences and projects, and the people I work with are the most rewarding part of my career.

Q. Where do you see yourself in ten years?

A. I'll probably stay in industry. I like the challenges in building actual things that are shipped to actual customers.

Q. Do you have any favorite books or resources that you'd recommend?

A. There is a small booklet I came across recently called Radical Careering by Sally Hogshead. It is quite inspiring. The other book I read recently that fits into the career management category is Secrets to Power Salary Negotiating by Roger Dawson. I would encourage everyone, and especially women, to be proactive in their career management.

Q. Do you have any quotes or mottos that you like?

A. When I was a girl, we had small albums where we would ask our friends and family to write little quotes or things for us to remember. I once asked my dad to write something in mine, and on the first page he wrote, "Live each day as if it were your first and last day." This has stuck with me for many years. Each day I try to look at things with fresh eyes. The future is wide open and the sky is the limit. This gives me permission to question everything, to start over, to get a fresh start, a second chance. At the same time it helps me prioritize by asking the question: "What would you do if you had only this day left?" Reminding us of our mortality also brings the "crisis" at work, or the small daily annoyances into

RELIABILITY ENGINEER

perspective.

Q. That's a great perspective. Is there anything else you would to add which might help fellow scientists who want to be successful in industry?

A. It's important to realize that every company is in the business of making money, and ideas are only good if they make money. You also have to look out for your interests and actively plan, work for, and negotiate advancement and appropriate compensation. You have to constantly update your technical skills and expand your network. You have to find out what employers are looking for and where the industry is going. Graduation day is not the end of learning, but your first day on the job.

TANJA BESHEAR

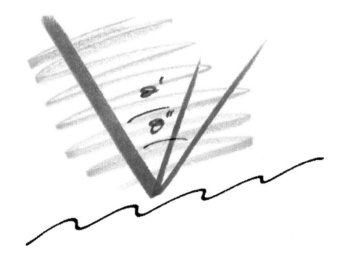

"Failure is simply the opportunity to begin again, this time more intelligently."

- Henry Ford

5

"One of the greatest things about a science Ph.D. education is that you are constantly learning to solve problems you haven't even encountered before. It's incredibly resilient training and very useful for an entrepreneur."

- Peter Fiske, Ph.D.

Peter S. Fiske is the CEO at PAX Water Technologies, Inc. in San Rafael, California. A native of Bethesda, Maryland, Peter has an A.B. from Princeton University, a Ph.D. in Geological and Environmental Sciences from Stanford University, and an MBA from the University of California, Berkeley's Haas School of Business. In 1996, he was selected as a White House Fellow and served one year in the Pentagon as Special Assistant to the Under Secretary for Acquisition and Technology, Dr. Paul Kaminski. In 2001, Peter helped found RAPT Industries, to commercialize a novel surface processing technology developed at Lawrence Livermore Laboratories.

In addition to his day job, Peter is a nationally recognized lecturer on the subject of career development for scientists and engineers and author of the book <u>Put Your Science to Work: The Take-Charge Career Guide for Scientists</u>. His columns have appeared in <u>Nature</u> and the on-line version of <u>Science</u>.

INTERVIEW

Q. Tell me about the science phase of your career. What first interested you in science?

A. I was born into science. My father is a geoscientist and he would take the family to the mountains of California each summer to do field work. From a very early age I was literally living science, and grew up with a very romantic and wonderful notion of what science was like. I was more of an

PETER FISKE

extrovert than the typical scientist and not particularly gifted in mathematics. However, science looked like such a wonderful and exciting thing – the whole combination of exploration and discovery and learning – that I decided to follow in my father's footsteps.

I went to college at Princeton University and majored in geology, just like my father. I continued in the geosciences in graduate school at Stanford, and it was there that I encountered the real business of doing science. It was a great experience, but I knew at that point that I likely wasn't headed down the traditional academic career path. When I finished school, I took a postdoc at Lawrence Livermore Labs, where I had established a collaboration during grad school, and switched to planetary science and condensed matter physics.

By the end of my postdoc, I decided that I wanted to pursue a career in science policy, so I applied for and got a White House Fellowship. This is a program that recruits early to mid-career people from all walks of life to work for a year as a special assistant to the President, Vice President, or a cabinet secretary. I headed off to D.C. to start my new career, but after two months I realized I had made a terrible mistake. The pace of progress in the federal government was just too slow. I soon realized I wanted something a lot more entrepreneurial, so after a great year as a White House fellow, I went back and got a staff position at Lawrence Livermore Labs.

> *"One thing you learn very quickly in business is just how much you don't know, and how much you need to rely on other people."*

Q. How did you end up making the transition from the government research lab into industry?

A. I stayed at Livermore for four years, but during that time I

CEO - PAX WATER TECHNOLOGIES, INC.

was also poking around in entrepreneurship. I enrolled in the evening MBA program at the UC Berkley School of Business, and in my second year started fishing around for ideas for Berkley's business plan competition. I had this great idea that I would be an agent to link technologies at Livermore Labs to the Business School at UC Berkley. I went around the lab and gave a series of lunchtime talks about entrepreneurship and the business plan competition and asked anyone who was interested to give me a call. I got around eight calls from people. Four were pretty good, and one stood out. The one that stood out was an optics manufacturing technology that had essentially been abandoned by the lab because it was too high risk.

I worked with the inventor and we decided to found RAPT Industries. Then I wrote the business plan and entered it in the business plan competition. I still remember vividly the night they announced the winners. The first thought in my mind when they announced that we had won first place was, "Oh @#%&! I am going to have to quit my job!"

So, I quit my job and started working full-time to get RAPT up and running. We took in a small round of initial investments, and then leveraged that with a lot of government R&D funds. That is what I did for the next six years.

Q. That's a great story of your transition from research scientist to entrepreneur. Do you have any thoughts about the challenges of making that transition?

A. I feel that becoming an entrepreneur was very synergistic with my research experience. One of the greatest things about a science Ph.D. education is that you are constantly learning to solve problems you haven't even encountered before. It's incredibly resilient training and very useful for an entrepreneur. A real breakthrough moment happened for me at RAPT when our original investor ran out of money. I suddenly had to hustle to an extent that I never had before.

PETER FISKE

It was an extremely trying experience, but in the end left me feeling so much more resourceful.

Also in grad school you learn how to live on $1500 a month and after five years in grad school you get pretty lean. Contrast that with these poor business school kids who work at Morgan Stanley for a couple years and then want to become entrepreneurs. After a couple years they are having their shirts pressed every morning and they've got their time-share in the Hamptons. They've become dependent on a high salary. Nobody goes into a science career with the idea of getting rich. You go into science because you are passionate about a subject and because you love learning, discovery, and the idea of creating something new. Those same passions are at the heart of entrepreneurship.

Q. What is your biggest career achievement so far?

A. My biggest achievement is stepping into the leadership role at PAX Water and bringing it from the brink of failure to the point of possibility. This accomplishment required many skills, but it also included a surprising amount of science. One of the problems was that the company was trying to sell a product without a strong technical marketing message. I helped assemble relevant data and create a message that allowed us to sell a technical product to a technical community in a way that was rigorous and credible.

Q. So, what are the non-technical skills that enabled that achievement?

A. One thing that is critical in business is just listening to others. This practice led to a very important breakthrough at PAX Water. Listening is sometimes a challenge for scientists who think they always need to be the expert. What I've found in business is that customers will often tell you what they want. They may not quite understand what they need, but they will tell you what they want. All you have to do is sit there and listen and ask thoughtful questions, and you can get a good sense for where you need to go.

CEO - PAX WATER TECHNOLOGIES, INC.

Q. I am curious how your perspective has changed since you founded RAPT. Is there anything that you wish you had known when you started?

A. That's kind of painful to think about [laughs]. There's no better place to plumb the depths of your incompetence than in a startup. Every day you are saying, "Oh God! I don't know anything about that!"

Early on I thought that great technology – having a better invention or widget – was the key to a successful venture. I now appreciate that there are so many other things that are important, including having the right people around you. If you have a great technology and an imperfect team, you are in serious danger, but even an imperfect technology in the hands of the right team can be turned into an opportunity. That's the biggest lesson that I've learned, or more accurately, the lesson that I am still learning.

Q. What is the most rewarding part of your career?

A. The most rewarding thing for me is having the ability to create new things that have a life and grow on their own, be they ventures, technologies, or my books. Working with great people is also very satisfying for me.

Q. So where do you see yourself in 10 years?

A. I would like to keep doing what I am doing now. I want to work with teams that will create a number of new ventures and introduce new technologies that will transform our economy and our planet. I believe early stage technology entrepreneurship is absolutely pivotal for economic growth and for progress on difficult issues like climate change. It's something I feel very passionate about.

Q. In addition to your day job, you advise early career scientists who are working in industry. Can you describe this part of your career?

PETER FISKE

A. I give a two-hour career development workshop 12 to 15 times per year, typically at leading research universities and sometimes at professional society meetings. I also write an occasional column for a journal such as Nature or Science. Everything I do on the career side is an after hours gig, basically a hobby. I do all sorts of juggling with my schedule to combine my business and speaking travel, but it comes at the expense of less time at home. This is a topic I care deeply about, and I haven't heard anyone else out there speaking from the same perspective that I have.

Q. What advice would you give to a scientist working in industry?

A. One of the real hang-ups that scientists and engineers have is they feel like they have to be the expert and be able to provide all the answers. One thing you learn very quickly in business is just how much you don't know, and how much you need to rely on other people. The sooner that you learn that, the more successful you will be. I've found people to be wonderful about giving you the answer or guiding you to others who can help you.

> *"If you have a great technology and an imperfect team, you are in serious danger, but even an imperfect technology in the hands of the right team can be turned into an opportunity"*

Scientists come from a culture where technical accuracy is absolutely paramount. It completely frames all discussion. When they move into an industry environment they see teams implementing imperfect technical solutions and making imperfect choices, and they simply don't understand. Their natural tendency is to say the problem needs more study, but this is often impractical. The whole world of business is about tradeoffs and the time value of making a decision now versus later. The fact is, at a certain point you have

CEO - PAX WATER TECHNOLOGIES, INC.

to make a decision based on incomplete data and move on.

Scientists can also identify with their discipline and see themselves only as a chemist or physicist. They lock themselves into very specific roles within a company instead of thinking "I am a problem solver. I am resourceful. I can do a broader range of things." I am surprised that scientists rarely spend time seeking out and learning about other aspects of the business. They should get out of their hole and see what is happening in other parts of the business. They should spend half a day with the marketing group, for example. When they don't do this they risk slipping into a degree of arrogance where they think the R&D team is creating all the value for the company and everybody else is just fumbling around. By the time someone gets to that point, they are in deep trouble.

Q. Are there areas where science education could be improved to help students transition into an industry setting?

A. There are a number of things that frustrate me about graduate school training in both the sciences and in engineering. One is that a lot of the critical skills that scientists and engineers really need, such as leadership and communication skills, are not focused on in graduate degree programs. Universities seem to feel that students either come with those skills or they pick them up on their own, so there's no reason to train them. These soft skills can and should be trained, though. The science and engineering disciplines would be much stronger if we had more professional focus on developing people, instead of just technical skills.

Q. I agree completely. Are there any books or other resources you would recommend for scientists working in industry or as entrepreneurs?

A. One book that I like a lot is <u>Crucial Conversations</u> by Kerry Patterson. It's a book and training module, and I've found the whole protocol to be very useful for difficult and tense

PETER FISKE

high-stakes conversations. John Gardner's book <u>Self Renewal</u>, written back in the 70's, is also a fabulous book.

Q. Let's not forget your own book, <u>Put Your Science to Work</u>.

A. My book is ideal for people who are at the earliest stages of their career. It contains the things that I wish somebody had told me when I was in graduate school. It's my attempt to give back to the scientific community and the publication that I am most proud of, because I feel it has the potential for real lasting impact.

This is a topic I care deeply about, but not many people in our community are concerned enough to create some level of discourse about it. It is sad, because scientists offer so much and yet it's shocking how little of a role they play in society at large.

> *"There's no better place to plumb the depths of your incompetence than in a startup."*

Q. Is there any saying or quote that you particularly like?

A. "The early bird may get the worm, but it's the second mouse that gets the cheese."

Q. That's a great one. How does that apply to what you've talked about today?

A. I have seen it happen many times, where a great new technology is introduced but the non-technical traps and pitfalls prevent the inventors from being successful. The second team to attempt it has an advantage, because they see what plagued the first group. My advice to any technologist is to study the history of an idea, technology, or business plan, and be very thoughtful about what pitfalls have befallen it in the past. It's easy to think that it's all about the technology and get caught in the mousetrap.

6

"I appreciate people who have an attitude of getting in there, rolling up their sleeves, and making something happen. They will make mistakes, but they back up and try again."

- Tom Baur, M.S.

Tom Baur is the founder of Meadowlark Optics, Inc., a company founded in 1979 that designs, manufactures and sells precision optics for polarization control. He is also the founder of Team Baur, a company that is developing a 75-acre business park in Frederick, Colorado. Prior to founding these companies Tom was a research scientist at the National Center for Atmospheric Research (NCAR) in Boulder, Colorado. He has a B.S. in astronomy from the University of Michigan and an M.S. in astrophysics from the University of Colorado at Boulder.

Tom is the winner of the 2005 Bravo! Entrepreneur Award for Northern Colorado, and a member of the Graduate School Advisory Council for the University of Colorado. He lives on a ranch on the Pawnee National Grassland in northeastern Colorado, where he and his wife Jeanne raise Black Angus cattle.

INTERVIEW

Q. Tell me about the science phase of your career. What influenced you to pursue a career in science?

A. I was in the 7th grade when Sputnik went up. That got me interested in astronomy. In high school I built a telescope and did quite a bit of astrophotography. My work in that area took me to the National Science Fair, and that gave me some encouragement to continue. I studied chemistry and astronomy at the University of Michigan, and then earned a master's degree in astrophysics from CU-Boulder.

After graduating I taught at a small liberal arts college in Illinois for a couple years and then joined the High Altitude Observatory, which is part of NCAR in Boulder. I was there about 13 years, doing quite a bit of observational astronomy and instrumentation development for solar physics experiments.

Q. How did you make the shift from scientist to entrepreneur?

A. I grew up in farming, which left me with a strong focus on practical applications, not theoretical pursuits. The instrument development I did at NCAR formed a bridge into an engineering role. However, there were a couple of specific things that influenced me to start my own company. I have a strong personal need for challenge, and I wasn't able to satisfy that need at NCAR. I found that I couldn't take on many of the challenging tasks or work independently, because I didn't have a Ph.D. One year I won the NCAR Technology Advancement Award, which was given out annually to one of the 500 scientists in the organization. I thought that would open some doors for me, but instead it seemed to create more jealously in the organization. At that point I decided I had to do something different, so I started Meadowlark Optics in my spare bedroom.

Starting my own company was all about my need for risk and challenge. I could not satisfy that need in a structured research environment where credentials seemed to be more important than achievement. I also felt that the link between performance and funding was not clear. What you accomplished seemed to matter less than the current political environment. I need more of a connection between risk and reward. I also need to feel useful, and

> "I need more of a connection between risk and reward. I also need to feel useful, and that was lacking for me in pure research."

that was lacking for me in pure research. The link between mapping magnetic fields in solar active regions and helping mankind was too loose for me.

Q. Did starting your own company fulfill those needs?
A. Yes. It's been pretty scary at times, but very satisfying.

Q. What would you consider your biggest achievement along the way?
A. I hate to say survival, but over the past thirty years that has been important, and not easy to achieve. I began Meadowlark as a boot-strap operation and we still have not taken any external money. We've developed a line of products that are useful to people and that's important to me. I am also proud that I've provided employment for lots of people, and helped them grow in their careers.

Q. What are some of the skills that allowed you to survive?
A. It's very important to hire the right people. Building a company is a tremendous amount of hard work. I was fortunate to not know what I was getting into. That reminds me of what Alexander Graham Bell said about developing the telephone: if he had understood more about what he was trying to do, he never would have attempted it.

For me, starting a company has been about the opportunity to learn new things. I'm not very good at repeating similar tasks over and over. I like to take on new technical challenges and be on the front end creative side of the work. I like to talk to customers and learn about their needs and where that overlaps with our technology. I enjoy trying to find new products that fit our expertise. These are fun puzzles for me to work on.

I also enjoyed learning about business. I see technical people who fail at operating a business because they don't respect the knowledge base that is required. You have to learn these things yourself or you've got to put people around you who know what to do.

Q. How did you solve that problem? Did you learn it yourself?
A. Yes, at the beginning I did, but what I found is that it became less interesting as I understood more, and there was less to learn. Now I hire other people to make the business machinery run so I can work on technical problems again.

Q. What is your role with the company now?
A. My title is "founder," but I'm the person who pushes the technology into new areas. I get us into the crazy projects.

Q. Have you hired other scientists into your company?
A. We have hired quite a few Ph.D. scientists over the years, but we generally don't have as much success with them as we do with masters and bachelors trained people. A lot of what we do is custom design work. In that environment, you often run into interesting technical problems, but you don't have time to study them completely because you've got a customer waiting. The employees with Ph.D.s seem to be more interested in chasing an interesting problem until it is fully understood, which is just what they had to do to complete their degree. It's hard to have the discipline to turn your back on an interesting problem and focus on the customer's needs instead. In some cases you may never understand exactly why what you are doing is working. That's a difficult thing for inquisitive minds.

Q. What are the characteristics of your more successful hires?
A. The people who are successful are good communicators, work well in teams, enjoy helping both internal and external customers, and are willing to put aside their intellectual curiosity when necessary.

Every now and then someone who is looking for work will come to me and say, "I have this really great idea," and I say,

FOUNDER - MEADOWLARK OPTICS, INC.

"Get outta here! I don't need any more great ideas. What I need are people who can execute." Success in a small company depends less on technical prowess than it does on many other skills.

Q. What were some of your biggest challenges in becoming an entrepreneur?

A. I didn't have a clue about the market for our polarization technology. Gaining that knowledge was a huge challenge. Learning about business was also challenging. To this day I have never worked for a for-profit company other than this one, so I had to make it up as I went along.

Q. What role does ranching play in your career?

A. Ranching provides the variety that I need. When I'm out on my ranch, my mind is blank and I have an opportunity to think more creatively. A lot of my best ideas come when I'm chasing cows.

Q. Very interesting. How do you split your time between your ranch and Meadowlark?

A. I spend about 90% of my time at Meadowlark and 50% at the ranch [Laughs].

Q. How has your perspective changed since you started Meadowlark? Is there anything you would do differently if you were to start another company?

A. That's a very good question. I don't want to suggest that we didn't make any mistakes, but I don't know that I would do a lot differently. I would probably try to boot-strap again. I'm not comfortable asking other people for money. I think its part of my agriculture background; you just don't do that.

As for a change in perspective, I have learned to curb my enthusiasm and not be so easily swayed by customers who claim, "If you can just make one of these for me today, I'll buy a million of them next year."

Tom Baur

Q. Where do you see yourself in ten years?

A. God, I hope I am retired! [Laughing]. No, I enjoy what I am doing. I like to work and in ten years I expect to still be working. What I want is more control over my own time, more freedom to split my time between Meadowlark and the ranch.

Q. You began your career teaching. Do you ever think about going back to that?

A. No, I don't think I'll ever go back to teaching at a school, but I do enjoy helping people grow in their careers here at Meadowlark.

Q. Helping people grow is a big part of teaching.

A. Yes, it is, and it is hard work to do it well. Teaching people in a company is more like a lab experiment than a lecture class. Some of what I teach is technical, but a lot is about the discipline of product development, project management, attention to detail, and all those things we don't learn in school. You can get a 90% on your test and still that's pretty good, but when you send a product out the door, it better be 100%.

Q. Do you have any thoughts on what universities can do to better prepare science students for careers in industry?

A. There is a growing understanding among universities that most of their students will go into industry. I know that the University of Colorado at Boulder is putting an increased emphasis on team projects and communication skills. I was involved with a committee that explored this for the Optical Science and Engineering Program (OSEP) at CU-Boulder. We developed the 'Employable Quintet,' which are the five skills that a student needs to be successful in industry and a major focus of the OSEP. These skills include: in-depth technical knowledge, problem solving ability in a laboratory setting, flexibility in learning and working, the ability to work in teams, and clarity in oral and written communication.

People who come out of school with a science degree should look at going into industry as a new learning opportunity. Learning is what science is all about. These students have shown that they are good at learning and they shouldn't stop. They ought to be sufficiently diverse, and not look down their nose at learning about the business skills that can make science useful to society.

> *"Success in a small company depends less on technical prowess than it does on many other skills."*

The traditional science career is a linear progression from graduate student to post doc to assistant professor to associate professor to full professor with tenure. That's a very structured environment and pretty narrow, small world. It's certainly the right environment for some very brilliant people, but the learning has stopped for some academics.

Q. Is there a saying or quote that you like?
A. Yes, I have one by Teddy Roosevelt on my computer:

"It is not the critic who counts; not the man who points out how the strong man stumbles, or where the doer of deeds could have done them better. The credit belongs to the man who is actually in the arena, whose face is marred by dust and sweat and blood, who strives valiantly; who errs and comes short again and again, because there is no effort without error and shortcomings, but who does actually strive to do the deed; who knows the great enthusiasms, the great devotions, who spends himself in a worthy cause; who at the best knows in the end the triumph of high achievement and who at the worst, if he fails, at least he fails while daring greatly, so that his place shall never be with those cold and timid souls who know neither victory nor defeat."

I appreciate people who have an attitude of getting in there,

TOM BAUR

rolling up their sleeves, and making something happen. They will make mistakes, but they back up and try again.

FOUNDER - MEADOWLARK OPTICS, INC.

7

"Work quickly and figure out what doesn't work, so you can find out what does."

Jason Ensher, Ph.D.

Jason Ensher is a Senior Optical Engineer at InPhase Technologies. He earned a B.S. in physics from the State University of New York at Buffalo. He earned his Ph.D. in physics from the University of Colorado at Boulder, where he worked with Dr. Eric Cornell, who won the 2001 Nobel Prize for demonstrating Bose-Einstein Condensation. After completing his Ph.D., Jason went on to a postdoc at the University of Connecticut with Dr. Edward Eyler, and then began his career in industry. Since then he has worked at ILX Lightwave, Precision Photonics Corporation, Ball Aerospace, and InPhase Technologies, all in the Boulder, Colorado area. In his spare time, Jason and his wife enjoy skiing, hiking, and swing dancing.

I met Jason when I came to work at Ball Aerospace in Boulder. Despite the fact that we worked in different areas during our time at Ball, our similar educational backgrounds provided us with good conversations. We've stayed in touch long after moving on from Ball, and frequently discuss our career experiences and plans. I am pleased to include Jason's input here.

INTERVIEW

Q. Can you tell me about the science phase of your career?

A. I developed an interest in optics and quantum mechanics by my senior year of college, so I applied to the graduate program at CU-Boulder. They had a large number of faculty in atomic, optical, and laser physics. They seemed to have a great program, and I liked the fact that Eric Cornell and Carl Weiman were working to achieve Bose-Einstein Condensation (BEC). I was also interested in living in Boulder. Eric was building his group at the time and needed some help over the summer, so I started working in his lab right away, studying ultra-cold atoms. By June of 1995 we had succeeded in achieving BEC. After I graduated I decided to take a postdoc position at the University of Connecticut where I worked on creating ultra-cold molecules out of ultra-cold atoms.

Q. So after an excellent start in academia working on research that earned your advisor the Nobel Prize, you decided to go into industry? Can you describe what was behind this decision?

A. During my postdoc I began to feel a little dissatisfied. I decided that maybe a tenure track position was not going to be fulfilling for me. I was doing interesting work, but it didn't feel like I was reaching enough people with my work. I wanted to do something that would have an impact on a lot of people, not just the several dozen colleagues around the world who might be reading my papers. Also, working with Eric and Carl in graduate school, I learned how to be successful in research, but I saw the investment that it requires. I decided that I was not wiling to commit at that level. I wanted a little bit more out of the rest of my life.

At the same time, fellow students who had graduated were reporting general satisfaction working in industry. I remember receiving an email from a friend who was working at 3M, and he indicated he had discovered this remarkable thing called Saturday. In the labs in graduate school, we typically worked much of the day on Saturday. Industry also provided higher compensation and a tangible reward, because you produce things that people use.

> *"...it is very important to be able to talk intelligently about the work you are doing."*

Through our work on BEC, I learned that doing something

new and risky can really pay off. I took that mentality with me to industry. I liked the idea of a smaller company that was charging ahead into the unknown, doing something novel and interesting. My first job was at ILX Lightwave. Telecommunications was growing rapidly at that time, and so was ILX. They had just opened a small development lab in Boulder, and there were only two other Ph.D.-level physicists there, so it seemed like a place I could carve out a niche for myself.

Q. How did you find industry to be different than your experience in academia?

A. One difference is the pool of engineers you have to learn from. In industry there are mechanical, electrical, and possibly optical engineers, who are well trained in their disciplines. You can work closely with them and ask them questions. At a university there may be a machine shop and, if you are lucky, an electronics lab. Other than that you are on your own.

I've also seen the difference between academic research and product development. In academia, the product is the science itself and the hardware doesn't have to be so robust. It is important that it give accurate data, and precision measurements may require stability over a matter of hours or days, but it doesn't need to operate longer than that without adjustment.

One of my favorite comments from Carl is that the perfect experiment set-up gives you your last measurement and then falls apart, indicating that you put just enough effort into it to get your measurements and didn't waste time making it too perfect. This is very different from developing a product that needs to work for a long time without continuous adjustment by a graduate student.

JASON ENSHER

Q. What did you learn in school that helped you make this transition?

A. Above all I learned the value of working quickly. My advisor used to say "Any job worth doing well is worth doing fast." I have had people debate that with me, but if you get the result you want, any more time spent on it is wasted. This also pushes you to explore the parameter space of failure, because no matter how smart you are, you are more likely to fail than you are to succeed. Work quickly and figure out what doesn't work, so you can find out what does.

I came to industry with a questioning attitude, more so than the other engineers who had been working there for a long time. I questioned what the product had to do and what the development effort had to achieve. I asked the market-

> *"My experience is that most scientists can benefit from learning more people skills."*

ing people about what the customers wanted. If one of the engineers indicated that we needed to build a circuit to do something, I would ask why. I wanted to know that we were being efficient and working as quickly as we could.

Second, being able to demonstrate working hardware is so important, even if it is not working perfectly. It is often necessary to demonstrate some basic level of functionality in order to get management to fund the next stage of a project, or to demonstrate capability to a customer. In graduate school I learned how to interpret 'bad' data, to be able to explain the effect of a certain piece of test hardware that wasn't working properly that day. This is exactly the kind of approach to take when demonstrating a prototype product. Having the ability to explain to your boss or a customer why it is not working to specification and what you can do to fix it is so much better than saying, "I don't have something that works for you now."

What has always gotten me the most traction in industry is being able to show data - hard facts, not just analysis. Theoretical analysis can be very important, but you can spend a lot of time on it. If you have the skill to perform an experiment in one day that demonstrates what you are talking about, you can save weeks of time. I have found that evidence almost always settles a dispute, even if the disagreement is political. Data-driven conclusions are very important.

Finally, I think it is very important to be able to talk intelligently about the work you are doing. My father was a technical salesman, and he emphasized to me that you need to be able to talk about what you are doing or you won't convince people how good your product is. I had great coaching on this in graduate school. The students would give talks in front of a big group and Eric and Carl would critique us, often on the language we used to describe our work. As a scientist and graduate student you tend to become comfortable talking about your ideas and having them challenged, and this skill is very useful in industry. I've seen many scientists who were successful in business development and marketing because they had these skills.

> *"Contacts and networking are important, because they are your best advertisement."*

JASON ENSHER

Q. Are there any areas that scientists may need to develop in order to be successful in industry?
A. My experience is that most scientists can benefit from learning more people skills. As scientists we need to know what we are talking about, but not come across as arrogant know-it-alls. Generally speaking, people hire folks they want to work with. You've got to be able to convince people that you are not threatening and that you are willing to learn from them. Part of this is the ability to discuss your ideas with people who may know little about your expertise, but are

very skilled at what they do.

One very important skill for me has been learning to work with and leverage the skills of other people, and not let a fear of sounding stupid keep me from asking for their help. I've found that physicists tend to develop a strong, overarching perspective on the product – to understand the whole picture, but they can learn a lot about the details from the design engineers. I also learned to invest in people and build both a technical and personal relationship with them. When you build these relationships and people trust you, they will put out a great effort for you when you are facing a challenge.

Q. What have been the most rewarding aspects of your career?

A. I have found it rewarding to work with teams of people, and to learn the leadership and organizational skills necessary to direct the activities of a team. I've also enjoyed the opportunity to work in so many different areas and learn about so many different things. I've worked at four different companies in a little under ten years - not by choice, because I was laid off twice - but I've worked in telecommunications, test and measurement hardware, optical component manufacturing, remote sensing, aerospace, and holographic storage. It's been rewarding to keep my career going and yet do these different and interesting things.

Q. You were laid off a couple of times during those ten years, but in each case you managed to find another job in only a couple of months. Can you outline how you sell yourself so effectively?

A. One of the most important factors was that I knew people where I was trying to go. Contacts and networking are important, because they are your best advertisement. Networking is key. That is how you find jobs in the first place, and it gives you a trusted voice within the organization. A

SENIOR OPTICAL ENGINEER

scientist can have the additional challenge of trying to get a job in industry. People may trust that you are a technical expert, but may question how you will perform in a practical setting. The very fact that I know people and use my network indicates that I have some fundamental people skills.

When I was first looking for a job in industry, I told interviewers that my graduate school experience would be useful in a manufacturing environment. I had spent five years developing a process for creating BEC, and then worked to optimize the process. This is what you do in a prototype and pilot production environment. So I said I had over five years of manufacturing experience. The fact that I was able to relate my experience to a manufacturing environment piqued their curiosity.

Q. Looking back over your career so far, is there anything you might do differently with the perspective you have now?

A. I might have approached things with a little more boldness. I might have taken the chance to start a business right out of graduate school, or right out of postdoc. Had I known more about the process, I might have said to my thesis or postdoc advisor, "There are some interesting things that we do here that might be valuable to other people." When you are younger and used to working for very little money, it is easier to do this, and a university provides a fertile ground for good ideas. As you get older, mortgages and other pressures begin to narrow your choices and make you a little less likely to take risks.

Q. Where do you see yourself in 10 years?

A. I would like to be in charge of a company that I start.

Q. Do you have any books or other resources that you would like to recommend?

A. One book that comes to mind is <u>Leadership Secrets of Attila the Hun</u>. I like history and this book is written in a very

JASON ENSHER

breezy style. In academia you never talk about what it takes to be a good leader and get people to follow you. I thought it was eye opening to think of Attila the Hun as an icon to be followed. This book had some interesting points about what people look for in a leader. It also addresses some of the hard things you have to do as a leader. Attila had to cut off a few heads to consolidate his leadership over the Huns. It made me think about how to address conflicts and establish my case.

There is a lot of passive-aggressive behavior in academia. You've got to be much better in industry, because there are people who have much better people skills than you do. What I see in VPs and senior managers in industry is something that is hard to put your finger on. You know it when you see it, it's emotional intelligence, emotional control, and self-discipline. It's that ability to control their own reactions, but also be able to read other people. The people who are put in charge are those whose judgment people trust. They are not going to lose their cool. They make data-driven decisions, but they also keep an open mind to how the customer, the business partner, the other engineers see the situation.

Senior Optical Engineer

8

"Do whatever excites you at the deepest level, whether that's being a professor, scientist/engineer, manager or an entrepreneur – or for that matter, an artist, daredevil, missionary, explorer, or anything else. If it doesn't excite you, you won't be happy and you will not excel."

Jack Jewell, Ph.D.

Jack Jewell sparked widespread industrial development of the Vertical Cavity Surface Emitting Laser (VCSEL), by demonstrating over 1 million VCSELs on a single chip at Bell Laboratories in 1989. Building on this achievement, in 1991 Jack co-founded Vixel Corp, the first company committed to VCSEL commercialization. In 1995 Jack left Vixel to start Picolight, Inc., and quickly established them as a leader in the commercialization of VCSELs and VCSEL-based transceivers for the data communications industry. With its world-leading VCSEL operation, Picolight was acquired by JDSU in 2007. Today VCSELs are used by billions of people each day in data communications and computer mouse applications.

Jack received his Ph.D. in optical sciences from the University of Arizona in 1984. He presently holds 68 U.S. patents, and has over 250 publications to his name. He is currently enjoying entrepreneurial activities, consulting, and running ultra-marathons.

INTERVIEW

Q. Tell me about the science phase of your career. Why did you originally decide to go into science?

A. I always liked science. When I started college, I knew physicists weren't in great demand, but I went into it anyway. It excited me so much more than any other field. I earned a master's in physics at Florida Institute of Technology (now Florida Tech) and then went to the University of Arizona (UA) to work on a Ph.D. After a year I switched to the College of Optical Sciences. They had a very strong school with a dynamic program. Many optics textbooks were written by UA professors, and their graduates were highly sought after in both industry and academia.

Q. Did a degree in optical sciences have more of an engineering focus than a degree in physics?

A. The department was pretty broad. They had everything from lens design, which is definitely an engineering discipline, to optical physics, which is fundamental science. My thesis was on optical bi-stable devices, which are switches made with optics rather than electronics. The primary application for these devices was in optical computing. This project had some engineering aspects because it involved building devices, but it also required material science and semiconductor physics as well.

Q. So you were still doing science research, but with an eye towards a specific application?

A. Yes. I wanted to make things that would make a difference in other's lives.

Q. Did you transition into industry right after school?

A. Yes, I went to work at Bell Labs right after school, but it was very similar to an academic environment. You were expected to give talks on your work and publish a lot of papers, but you didn't have to raise money. I actually continued working on the same topic as my Ph.D. thesis.

Bell Labs was a tough place to work. Some of your strongest competitors were inside the company. That helped us all keep more active, of course. When I look back at it now, though, after my experience in start up companies, my time

Founder - Picolight, Inc.

at Bell Labs seems like an extended childhood, because we didn't have to generate monetary support for our work.

Q. How did you transition from Bell Labs to starting your own company?

A. I spent my time at Bell Labs improving my optical bi-stable devices, and I was always interested in starting a company based on this work. After a few years, though, optical computing wasn't looking very promising as an application area. The devices that I made had a structure that was very similar to a laser, so I made a few changes and turned them into vertical cavity lasers. That led to a big breakthrough in VCSEL technology, and was a real turning point in my career.

Following that success I figured I would probably never have a better opportunity to start a company. I had been talking to other people about this, and we all agreed that this was something to go for. Because of Bell Labs' emphasis on publishing and presenting at conferences, my work was very well known, so it was relatively easy to get SBIR and other funding for the company. Two years after the breakthrough I left Bell Labs and did consulting work to keep the money flowing until our contracts kicked in. We started off as Photonics Research, Inc. and later renamed the company Vixel Corporation.

Q. Is that when you moved to Colorado?

A. Yes, I always wanted to move back to the western U.S. This is where I feel most at home. Two other people I started the company with were working at Sandia National Labs in Albuquerque. There was also a company called BandGap Technologies in Broomfield, Colorado that had Molecular Beam Epitaxy (MBE) and Metal-Organic Chemical Vapor Deposition (MOCVD) facilities that were available for us to use. They were interested in branching out into electro-optical devices, so there seemed to be a great match. After a while it became clear that they wanted to own us and hire us as

JACK JEWELL

employees, so we split off from them after a year.

Q. Are there skills you developed as a scientist that you found useful for starting your own company?

A. Absolutely. There are a lot of skills that transferred over. The writing skills I developed writing technical papers made it easier to write proposals. I had also filed some patents at Bell Labs, so I had experience working with the patent attorneys and understood that process well. Giving presentations is also a great skill to have because you have to give presentations to venture capitalist to sell your company. I was used to getting up in front of people and trying to convince them that what I was doing was the greatest thing on earth, and would make everyone's life better.

Obviously I had to start looking at marketing and the economic side of the business as well. That was new, but not all that hard. When we were applying for SBIR's it was much like what a professor would do when applying for grants and contracts.

Q. What aspects of starting a company did you find particularly challenging?

A. Relationships were a big challenge. In the beginning we had partnered with BandGap Technologies, and that quickly fell through. Then there were a lot of challenges working with the CEO at Vixel Corp. That's why I left Vixel in 1995 and started Picolight.

Q. Did you leverage any of the work you had done at Vixel to start Picolight?

A. My technical expertise and personal connections were retained, but otherwise I started over. I hunkered down in my basement for over two years, doing a little bit of consulting to keep the money flowing while I wrote quite a few patent applications and SBIR proposals.

Q. Were you making devices during those two years?

A. Yes, I leveraged a number of university collaborations to get the lab work done. I knew many of the professors in the field, so I gave them small subcontracts under my Phase I SBIRs to do experimental demonstrations for me. It made a lot more sense to have the work done at the universities than to try

> *"Many of the improvements I made were things that people didn't think would work. Having the guts to try anyway made all the difference."*

to build my own hardware at that early stage. It also gave the work more credibility and increased the proposal reviewer's confidence that we would make real progress during Phase II of the contract.

I didn't draw any salary from the company until the government contracts started kicking in. I had an angel investment, but I used that money for market surveys and other things that I couldn't spend SBIR money on. I wanted to get the best use I could from that money rather than pay myself. Companies don't start with money. They start with ideas and a lot of hard work.

By the summer of 1997, I had been approved for a few Phase II contracts, so I began to hire a few people to help. First I hired Stan Swirhun, who I had worked with at Vixel. He was extremely good technically, but he also had very good business skills. He had more experience with venture capital than I had, so he ultimately became the CEO and I became the CTO. I also hired a senior optics engineer and we had a temporary CFO who had been an angel investor in the company.

In January of 1998 we landed our first venture capital round. At the time we were all still working from spare bedrooms, but with venture capital the goal was to move into a facility

JACK JEWELL

69

TURNING SCIENCE INTO THINGS PEOPLE NEED

with a real lab and dedicated testing facilities. That is also when the real hiring started, and a year or so later we decided to install our own MOCVD reactor.

Q. What skills did you look for as you began hiring technical staff?

A. We would look for specific skills needed to execute our contracts and to build the company. We picked people both for their technical skills, and for their personal skills. We'd had enough of personality issues at Vixel and didn't want to have those problems with our new company. We had an advantage because of this other VCSEL company in town, and we knew all of the employees. One of our VC's called it "a hiring opportunity of epic proportion" [chuckles].

Q. Surviving the telecom bubble burst must have been a big challenge. How did you manage that?

A. Not easily. It was really, really difficult. We had to work just as hard, but instead of having all these amazing scenarios in front of us, it became a matter of just surviving. We ran out of money and we did a couple of down-rounds. The value of the company plummeted about 25 times.

Q. You managed to survive, and in 2007 you were acquired by JDSU. Are you happy with how that turned out?

A. It was a lot better than any other choice available at the time. If we had sold out during the bubble, that obviously would have been better. If we had completely run into the ground, that obviously would have been worse.

> *"Companies don't start with money. They start with ideas and a lot of hard work."*

My advice for others who are growing a start-up is that if you have a good exit opportunity, take it. I've told other people to do that. I was running a hundred mile race and chatting

FOUNDER - PICOLIGHT, INC.

with a younger guy who had a start-up company. He was considering a buyout opportunity that would force him to move to Los Angeles for a few years. I told him, "Just do it, and do your time. You'll have a success on your record and move onto other things. I did 7 years in NJ, so you can do two or three in LA." I heard later that he took my advice.

Q. What do you consider to be your biggest career accomplishment?

A. Without a doubt, the VCSEL demo I did at Bell Labs in 1989 is my biggest accomplishment. It transformed the VCSEL from a research curiosity into a practical device.

Q. What are the key factors that enabled this success?

A. One factor was the freedom I had to move in whatever direction I felt was appropriate. I had spent a lot of time engineering the devices to be small and fast. Many of the improvements I made were things that people didn't think would work. Having the guts to try anyway made all the difference.

Speed was important in order for the devices to be practical for optical computing, so the active region needed to be very small. We used a quantum well structure for the active material, and if you want to minimize the number of wells you need very high reflectivity mirrors. People had demonstrated mirrors with 90% reflectivity, but I set out to find out just how high a reflectivity we could achieve. We did some experiments that suggested you could reach higher than 99% with no limitations on device performance.

Next I set out to reduce the number of quantum wells. After a number of steps we demonstrated a device with a single quantum well. That was hard to do, but we just kept increasing the number of layers in the mirrors to get the reflectivity as high as we could. This demo used optical pumping, so the next step was to create an electrically pumped structure. That gave us new problems because the multi-layer mirror

Jack Jewell

structures were not very good electrical conductors. I saw a paper that talked about using a graded superlattice for a non-metal contact, so I decided to try adding this to each mirror interface. That ap-proach gave us what we needed and enabled the breakthrough demonstra-tion.

> *"I wanted to make things that would make a difference in other's lives."*

Q. So a key element of your success was having the guts to challenge the conventional wisdom?

A. Yes. People didn't think you could make mirrors with a high enough reflectivity, or send electrical current through such mirrors. Rather than assume a limitation, we pushed the technology as hard as we could to find its real limits.

Q. What has been the most rewarding aspect of your career to date?

A. Seeing widespread adoption of the VCSEL has been very re-warding. I had a video interview back in 1992, where I was asked, "What is your dream for your company?" I answered, "My dream is to walk into a consumer electronic store and see it full of our VCSEL products." We are getting there now. All of our email and Internet traffic travels through VCSELs, and if you have an optical mouse on your computer, you have a VCSEL in the palm of your hand. I can't wait to see them used in even more applications.

Q. So you are making a difference in peoples lives, just as you always wanted. Where do you see yourself in 10 years?

A. I have no idea. And if you had asked me that question ten years ago, I would have given you the same answer. I would like to do something different, but I haven't decided what that will be. The datacom industry has not been much fun, but after more than twenty years in the industry, it is easier to be smart in datacom than it is to be smart in something new.

FOUNDER - PICOLIGHT, INC.

Q. Do you have any other advice for a scientist working in industry or interested in becoming an entrepreneur?

A. Do whatever excites you at the deepest level, whether that's being a professor, scientist/engineer, manager or an entrepreneur – or for that matter, an artist, daredevil, missionary, explorer, or anything else. If it doesn't excite you, you won't be happy and you will not excel.

JACK JEWELL

"Luck is the residue of design."

- Branch Rickey

"Being able to face the truth – whether it's something you like or not – is important."

Ashok Balakrishnan, Ph.D.

Ashok Balakrishnan is the Director of Product Development at Enablence Technologies in Kanata, Ontario. Ashok was a co-founder of Enablence Inc., and is a co-inventor on the key patents that define Enablence's proprietary technology. He has extensive product development and commercialization expertise spanning several markets, including spectroscopy, telecommunications, and biophotonics. Prior to Enablence, Ashok was the manager of product integration at Optenia, a Staff Scientist at SDL (now JDS Uniphase), a Photonics Development Scientist with Mitel Semiconductor (now Zarlink). Ashok holds a Ph.D. in Physics from the University of Toronto, and was a postdoc researcher with the U.S. National Institute of Standards and Technology (NIST). In his spare time, Ashok enjoys running and spending time with his family.

I first met Ashok when I came to work at SDL in 1996. For two years we worked together in the Systems Group developing single-frequency tunable lasers for spectroscopy applications. It was great to reconnect with him and get his perspective on this topic.

INTERVIEW
Q. Tell me about the science phase of your career.
A. As a child, I was always interested in math and was good at it in school. I liked science as well, but math was always the most exciting. Then as I got into high school, I realized that I had a better chance of getting a science-related job, so that pushed me in the direction of science. As an undergradu-

ASHOK BALAKRISHNAN

ate, my degree was actually in engineering physics, which was basically an engineering degree with more emphasis on physics. Some classes taken with the physics students at the university.

After my undergraduate years, I was faced with the option of going right into the job market as an engineer or going on to graduate school. I looked at the opportunities that were available, but didn't find anything that I felt would challenge me at the level that I was looking for. I wanted to do something that involved a lot more high-end problem solving, so I made the decision to pursue science more deeply. I went into the graduate program at the University of Toronto where I earned both an M.S. and Ph.D. and performed research in laser spectroscopy.

> *"As a scientist you are not as exposed to the need for strong social skills, but I think these are one of the most important aspects of success in business."*

After my Ph.D., I did a two-year postdoc at the National Institute of Standards and Technology (NIST). I worked in very fundamental laser spectroscopy research. That was the end of the pure science part of my career.

Q. Tell me about your transition into industry.

A. One reason I decided to work in industry was that the project I was working on at NIST was fairly lengthy, and didn't get far enough during my two-year postdoc that I was ready to publish a paper about it. I realized this narrowed my options if I wanted to become a professor or carry on fundamental research. It limited the number of places I could look for a job, so I figured perhaps I should go into industry.

I also knew there were a lot of interesting problems to be solved in industry. What I found most exciting in academia

DIRECTOR OF PRODUCT DEVELOPMENT

was trying to figure out how to put the equipment together to get the results you needed. For me, designing the experiment was always more interesting than the answers that might result. At the time, I didn't know exactly what was involved in designing a product, but I knew it would include the design and assembly aspects that I found so interesting.

When I was a postdoc, I was using lasers from SDL, Inc. I thought it would be interesting to design lasers at a company like that, because of all the neat things they must have access to. I figured I could learn the answers to all of the questions I had about their lasers if I were able to work there. Luckily, SDL had an opening that was in line with what I was interested in. I got a job developing lasers for Raman spectroscopy that was a pretty good fit, and made a great transition into industry.

Q. What did you find to be the most challenging about the transition?

A. I was surprised to find that in product development there are a lot of tiny little problems that have to be solved. There's also the big picture. Of course you have to get the science and the fundamentals of the design correct, but even if that's right and everything looks elegant, there are all these little tiny things that have to be solved. You have to get the size of a screw right and there may be certain things you want to do but are too expensive. You might also have a problem finding a vendor that sells the right parts or is willing to make things the way you need them to be made. There are a lot of practical issues that you are faced with.

I was also shocked by the time pressure in industry, where a very long project might take a year and often much less. Projects take much longer in an academic research environment. A Ph.D. takes many years and you are expected to take your time, look at every single detail of your problem, and know it inside and out. In industry, it's the bottom line

ASHOK BALAKRISHNAN

that counts and it's okay if you don't understand every little detail of your product as long as it works properly. It is only when it doesn't work properly, that you invest the time to go back and understand it better.

Q. What would you consider your biggest career achievement?

A. In 2004 a few colleagues and I formed our own company. The company is called Enablance and that is where I'm working now. When we started the company, it was very small and just a way to stay employed and keep food on the table, but it has grown in surprising ways. Now it employs 25 people or so locally, and close to 200 people internationally.

Q. What would you say are the primary skills that enabled you to accomplish that achievement?

A. The most important was just knowing what is important and what's not. Whether you are building a product or setting up a company, there are certain things that are important and have to be done right now. Other things may seem important, but don't really matter. It is critical to be able to assess a lot of pressures, and then determine which ones are critical and which will not really matter in the end.

Being able to face the truth – whether it's something you like or not – is important. I learned this from graduate school. There were times in graduate school when I would struggle for months trying to figure out why I was not getting the results I was expecting. I would spend time pursuing problems I could fix more easily such as a mirror alignment or laser power, but all the time have this nagging feeling that something else was really the problem. It was something else that had been staring me in the face for a long time, but I was not willing to admit it. The moment you admit that, you can solve the problem.

Facing the truth is just as important in industry. Sometimes you are in the middle of developing a product and you real-

DIRECTOR OF PRODUCT DEVELOPMENT

ize, "Uh oh, this is not the way to go." You have to be honest with yourself and change direction quickly if you want to be successful, regardless of what you prefer. When you want to try to sell a particular business idea or product, you might think this is the niftiest thing in the world, but people may actually want something else. You have to deliver value, so be honest what that value is.

Q. How do you sell yourself and your skills?

A. When I was straight out of research I would say that I was very familiar with lasers and optics. Once I started working in industry I realized that it requires a familiarity with many disciplines. To develop laser systems you have to know optics, but you also have to understand mechanical design and thermal analysis. In order to make a product work, there are many different areas that you've got to understand. Early in your career you need to be a specialist in some area in order to get into a company or to get placed on a project. Later on, if you want to lead a project, you don't need to be a specialist, you just have to know enough about every area, so you can talk to the people with the expertise in mechanical design or thermal design or whatever.

After a year or two in industry, I realized there were other skills I had that were important. I used those skills to sell myself. Those included working well with other people, leading a technical team in a matrix environment where there is no direct reporting structure, and being able to get things done on time.

> *"It is critical to be able to assess a lot of pressures, and then determine which ones are critical and which will not really matter in the end."*

I found that in job interviews, people would look at my resume and think that because I had a Ph.D., I would be very research oriented or too much of a nerd. You do have to convince people that's not who

ASHOK BALAKRISHNAN

79

you are. You have to convince them that you can work with other people and you are interested in actually solving the problems that your employer faces, not just in solving problems for the sake of solving problems. You have to convince them that you are practical and are not going to take forever on a project. All of these things have to come out in an interview setting.

Q. Let's go back to the company you started. Have you hired both scientists and engineers? What are your thoughts on the differences?

A. In certain cases, we have hired engineers because we needed a specific skill set to solve a problem. In other cases we have hired scientists, because we were not quite sure what problems we might run into. We just knew that there was a general area that we needed help with. In this case we look for someone with a real spark, an interest in solving problems that no one else has ever solved. I find that if the problem we are trying to solve is something very tricky and high level, where we are delving into the unknown, scientists have an edge. If they have gotten a Ph.D. in physics, you know they are not afraid of the unknown, because that is the whole idea of the degree. They often actually thrive on discovering the unknown.

Q. I am curious how your perspective has changed now, compared to when you first went into industry.

A. As a scientist you are not as exposed to the need for strong social skills, but I think these are one of the most important aspects of success in business. There's a lot of social interaction involved and it has to be done right. You can develop a great product but nobody is going to buy it if you aren't good at selling to people. You may have a great company with great people and great ideas, but if you don't come across well to potential investors, then your company won't go anywhere.

DIRECTOR OF PRODUCT DEVELOPMENT

Q. What has been the most rewarding aspect of your career to date?

A. The most rewarding part of my career has been growing Enablence to the point where it has an impact on people's lives. There are people who have jobs because of my efforts. That gives me a good feeling.

Q. Do you have any saying or quote that you particularly like?

A. In my 11th grade physics class, when we would complain we couldn't solve some problem, my teacher would say, "Well, nature has a solution, so why can't you find it?" I think back to that periodically when I'm trying to figure out how something works. If it works in nature, someone had to have thought out an answer. Why can't I find that? Of course, working in high-tech you are doing things that aren't done in nature. Maybe the analogue for industry is, "Well, our competitors have a solution."

Ashok Balakrishnan

81

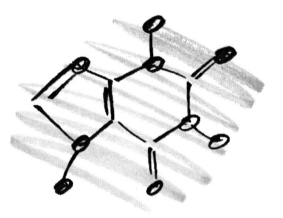

"Face reality as it is, not as it was or as you wish it to be."

\- Jack Welch

10

"One of the main challenges I faced in industry was switching from science research to product development. These two activities are very different."

Antoine Daridon, Ph.D.

Antoine Daridon is a Product Manager at SpinX Technologies in Geneva, Switzerland. He is also a member of the Strategic Advisory Board of InSphero in Zürich, Switzerland. Antoine holds a Ph.D. in analytical chemistry from the Universität Bern in Switzerland, a masters in analytical chemistry from the Université de Jussieu, Paris VI and the ESPCI, Paris Tech, and a B.S. degree in organic chemistry from the Université de Bretagne Occidentale.

After earning his Ph.D., Antoine took a postdoc position with the group of Nico de Rooij at the Université de Neuchâtel, where he developed microfluidic systems for environmental research. This experience led to an R&D position with Fluidigm in San Francisco, California, where he developed microfluidic assays for research in protein crystallography, cell based assays, chemical synthesis, and genetic analysis. When he first joined Fluidigm, Antoine was the only person in this role who did not have an engineering background. He feels this was beneficial, as it helped him see the product line from the point of view of the customer and ask, "What scientific information can I generate with this device?"

INTERVIEW
Q. Tell me about the science phase of your career.
A. I was attracted to science at a fairly young age. I enjoyed mathematics, physics, chemistry, and biology, so in college I got a bachelor's degree in organic chemistry. After that I

went to the ESPCI ParisTech and earned a master's in analytical chemistry. After that I served my country in the Navy and then went to the University of Bern in Switzerland to earn a Ph.D. For my research I spent half of my time at the University of Bern and the other half at the IBM research laboratories in Rüschlikon, Switzerland.

My research was in electrochemistry, but involved a lot of physics. I was trying to create high temperature superconducting nano-structures using a scanning tunneling microscope (STM) tip. This project was very interesting, because IBM Rüschlikon researchers had just received Nobel prizes in physics for the invention of Scanning Tunneling Microscopy and the discovery of High Temperature Superconductors, and my research involved both topics.

Q. What influenced you to go into industry instead of staying on an academic path?

A. It seemed that professors spend more of their time looking for funding, rather than actually doing science. In industry, someone else is dealing with the money issues so the scientists can focus on their research. That's what drove me towards industry.

Q. How did you make the transition to industry?

A. I felt that my experience at that point had been too deep in fundamental research to allow me to transition directly into industry, so I took a postdoc position at the University of Neuchâtel where I worked with microfluidic systems. The research that I did there was much more oriented towards industry, and gave me the opportunity to learn some engineering skills. The connections that I made there led to my first job in industry, an R&D position with Fluidigm in South San Francisco, California.

Q. Would you describe the work you do in your current job?

A. SpinX Technologies makes instrumentation for drug discovery research. As a product manager, I work with our customers to adapt their custom drug assays to our platform and make sure the instrument that we are developing fits their needs. I also improve the graphical user interface to ensure that it is intuitive and works well for our customers.

> *"I recommend that people find a good role model."*

Q. Are most of your customers research scientists?

A. Yes, they are. We turn the customer's needs into product requirements, so a big part of the job is translating how a software engineer thinks into what biologists can understand and vice versa.

Q. That sounds like a great role for someone with a research background. What do you enjoy most about your job?

A. I really love biology and instrumentation and I enjoy working with people. I like when I have to put my brain cells to work trying to solve a new problem. That is what keeps me moving. I am not very interested in repetitive work.

Q. What are some of the challenges you've faced in making the transition from science into industry?

A. One of the main challenges I faced in industry was switching from science research to product development. These two activities are very different. There are many product development concepts that I had not been exposed to in research. For example, I had no idea what a product development cycle was. The concept of product requirements was also new. As a scientist, you tend to pursue the absolute best performance you can get out of your equipment. In product development, you have to define requirements that every instrument produced can meet, and this often requires a compromise.

Q. How has your perspective changed since you started working in industry?

ANTOINE DARIDON

A. I used to focus on achieving a better understanding of some problem. Now I focus on what I can do to help someone else.

Q. **That speaks to the fundamental difference between science and engineering.**

A. Science is about exploration and gaining new knowledge, while engineering is more about applying existing knowledge. A scientist studies a problem in order to understand how or why a system works, while an engineer looks for solutions to a given problem using existing knowledge.

This is a generalization, of course. I've seen people who were trained as engineers doing fundamental science and people trained as scientists who are now doing engineering. It's about where you place yourself, in a research lab or in a product development environment.

> *"I like when I have to put my brain cells to work trying to solve a new problem. That is what keeps me moving."*

In college, I had a professor who had a twin brother, and they had both gone to the same technical school in France. My professor continued on to an academic career while his twin brother went into industry. The academic's perspective was that his brother may earn twice as much, but he was not discovering anything new. The one in industry looked down on his academic brother, thinking that he did work that didn't serve anyone directly.

Q. **That's unfortunate, because both are very important. Academic research provides many of the ideas that are pursued in industry, and without commercialization, the value of science research would never be realized.**

A. Yes, the work you do in science might be helpful to people ten to twenty years from now, and in industry your work needs to be useful to people in a year. Both are valuable, but the time before an impact is seen is not the same.

Q. What would you consider your biggest career achievement?

A. When I worked at Fluidigm in California, I helped to develop a microfluidic device that was used to identify pancreatic cancer. The goal was to identify a few copies of mutated DNA in the middle of a pool of normal DNA strands, but the signal-to-noise was too low to do this easily. It was like looking for a needle in a haystack. We developed an approach where we chopped the samples into many small chambers and spread it over a large area so that the mutated DNA stood out. It was analogous to spreading the haystack over a football field so the needle is surrounded by only a few strands of hay and is easier to identify.

This device was just a concept when I ran across it, and was not being considered for serious development. I saw that it would bring a new capability, so I invested some of my personal time to develop it further and show that it would work. Convincing the company to pursue this project is definitely my biggest achievement.

Q. What are the key skills that enabled this accomplishment?

A. Creativity, focus, and rigorous experimentation. Those are my main strengths.

Q. One of the challenges scientists may face is how to sell themselves in an industry environment. How do you sell yourself and your skills?

A. Selling myself has been a challenge, and I'm not sure I do it well. I tend to assume that those around me are clever enough to figure out on their own that I am competent and capable. That is naïve. In addition to working on my projects, I need to put effort into moving my own career forward.
When I do sell myself, I tend to give examples of what I

ANTOINE DARIDON

have achieved in the past to demonstrate I am creative. I talk about my achievements and the products I've worked on, and show how I am able to take a scientific problem and make it work in an industrial environment. I also talk about my management skills and how I enjoy working with people. I don't focus on my education, because the research I did in school is not very applicable to industry problems.

Q. Where do you see yourself in 10 years?
A. I would like to be an R&D director, focused more on product development than on early stage research.

Q. Are there any books or other resources you would like to recommend?
A. Two books that I found particularly interesting are <u>The Goal</u> and <u>Critical Chain</u> by Eliyahu Goldratt. <u>The Goal</u> is a novel that presents the "Theory of Constraints" as a method for performance improvement. <u>Critical Chain</u> discusses the application of this theory to project management. These books gave me a different perspective on the process of turning a technology into a manufactured product.

Another book I enjoyed, <u>Losing my Virginity</u>, an autobiography by Virgin Group founder Richard Branson, talks about the passion and focus needed to create a successful business.

Q. Is there anything else you would like to share?
A. I recommend that people find a good role model. My Ph.D. advisor, Prof. Hans Siegenthaler in Bern, has been a great role model for me. He is retired now, but I am still in contact with him. He has fantastic scientific knowledge and a great capacity for teaching, but he also has great human qualities. He was the person who welcomed me to grad school, gave me the tools to do my work, and made sure everything was going well from a personal perspective. When I am gone, I hope people will think about me the way I think about him.

PRODUCT MANAGER

ABOUT THE AUTHOR

David Giltner is a laser scientist turned photonics engineer, as well as a speaker, musician, outdoor enthusiast and father. He has a Ph.D. in physics and develops laser-based products for a variety of applications. Late in his graduate school career, he decide not to follow the 'traditional' path into academia and has instead found a very rewarding career "turning science into things people need". By compiling wisdom from other successful scientists working in industry, David intends to help others who are following a similar career path. He can be reached at **david@50interviews.com**.

Visit David's blog at **http://scientists.50interviews.com** and sign up to receive information on his forthcoming book of the same title.

ABOUT 50 INTERVIEWS

Imagine a university where each student not only gets a textbook custom tailored to curriculum they personally designed, but where each student literally becomes the author!

The mission of 50 Interviews, Inc. is to provide aspiring, passionate, driven people a framework to achieve their dreams of becoming that which they aspire to be. Learning what it takes to be the best in your field directly from those who have already succeeded. The ideal author is someone who desires to be a recognized expert in their field. You will be part of a community of authors who share your passion and who have learned first-hand how the 50 Interviews concept works. A form of extreme education, the process will transform you into that which you aspire to become.

Looking to dramatically alter the course of your life? First go have some conversations with those who do what you think you might like to emulate. Find out what drives them and how they found success. Then rewrite your future, one interview at a time...

50 Interviews is a career change process where you first fully explore potential future careers. BEFORE making any risky commitments, you first spend time talking to those who do the job you think you would love to try next, top professionals in their field! While gaining from their experience and knowledge, YOU gain great mentoring tips, phenomenal connections, and then share those with others in your own 50 Interviews title as the author of your own book!

50 Interviews is a showcase of people living their dreams, and enjoying the payoff that results from taking the leap to explore something they are truly passionate about!

Learn the truth about the challenges and the rewards of whatever field interests you. Gain valuable advice from people who were once in your shoes. If you don't make a change now, will you ever?

Sure you have doubts, but how else will you learn everything about pursuing that dream job? Your interviews will shed light on what has inspired and driven others to follow their dreams and how those dreams have manifested for them. It is a way to collect over 50 new mentors who might inspire you and prove that anything is possible, and then you can share that knowledge with the world through your own publication.

Each 50 Interviews book title is determined by the author who chooses the career they would like to explore in depth. By interviewing top professionals in that field, you gain a glimpse into the vital mindset of those successful in your field of choice, and then use that knowledge to take you one step closer to their own definition of success.

If you are interested in learning more, I would love
to hear from you! You can contact me via email at:
brian@50interviews.com, by phone: 970-215-1078 (Colorado),
or through our website:

www.50interviews.com

All my best,
Brian Schwartz
Authorpreneur and creator of *50 Interviews*

OTHER 50 INTERVIEWS TITLES

Additional topics based on the 50 Interviews model that have already been released or are in development:

Athletes over 50
By Don McGrath

Video Marketing Pioneers
By Randy Berry

Young Entrepreneurs
By Nick Tart and Nick Scheidies

Professional Speakers
By Laura Lee Carter & Brian Schwartz

Artists
By Maryann Swartz

Attraction Marketers
By Rob Christensen

Spiritualists
By Tuula Fai

Professional Speakers
By Laura Lee Carter and Brian Schwartz

Franchises
By Leslie Lautzenhiser

Wealth Managers
By Allen Duck

Millionaire Women
By Kirsten McCay-Smith

Entrepreneurs
By Brian Schwartz

Property Managers
By Michael Levy

Actors
By Stella Hannah

Reinvent yourself, one interview at a time.

www.50interviews.com

CPSIA information can be obtained at www.ICGtesting.com
Printed in the USA
BVOW030606060612

291907BV00003B/173/P